IRELAND'S LOST BIRDS

IRELAND'S LOST BIRDS

Gordon D'Arcy

FOUR COURTS PRESS

This book was set in 11 on 14 pt Garamond
by Carrigboy Typesetting Services, County Cork for
FOUR COURTS PRESS LTD
Fumbally Lane, Dublin 8, Ireland
Email: info@four-courts-press.ie
and in the United States for
FOUR COURTS PRESS
c/o ISBS, 5804 N.E. Hassalo Street, Portland, OR 97213.

A catalogue record for this title is available from the British Library.

ISBN 1–85182–529–0 pbk
ISBN 1–85182–526–6 limited edition

This publication has received support from the Heritage Council
under the 1999 Publications Grant Scheme.

AN
CHOMHAIRLE
OIDHREACHTA

THE
HERITAGE
COUNCIL

Printed in Ireland by
ßetaprint Ltd, Dublin

CONTENTS

LIST OF ILLUSTRATIONS

PLATES: SCRAPERBOARD ILLUSTRATIONS

FIGURES, PHOTOGRAPHS AND TABLES

CREDITS

All plates are by the author. With the exception of those otherwise credited, all figures and photographs are by Gordon D'Arcy. Figures 12, 16, 26, 30, 33, 39, 41 and 50 are courtesy of the Ulster Museum, Belfast; fig. 21 is courtesy of Cambridge University Library; and figs 37 and 42 are courtesy of Pierre Petit.

ACKNOWLEDGMENTS

I would like to thank the following who, in one way or another, helped in the preparation of this book. Due to its long gestation period some of those who contributed may have assumed more elevated status in the interim (apologies if this is indeed the case); one or two are unfortunately no longer with us.

Finbar McCormack and Professor J. Mallory of the Queen's University, Belfast; Professor Woodman, Margaret McCarthy, Tanya O'Sullivan and Dr Paddy Sleeman, of University College, Cork; Professors James Fairley, G. Mhic Eoin, Brendan Ó Madagáin; Dr P. Gosling, Jim Higgins, Michael McGrath, Áras na Gaeilge, NUI, Galway; Dr J.H. Andrews and Bernard Meehan, Trinity College, Dublin; Dr Pat Wallace, Nigel Monaghan, Ivor Harkin and Ned Kelly, National Museum, Dublin; David Irwin and Marshall McKee, Ulster Museum, Belfast; Professor T. de Bháldraithe, Dr B. MacShamhrain, E. O'Hogan, Marin Mac Ghadhadhonn, Royal Irish Academy, Dublin; Dr M. Speight, Amhlaoibh Ó Aonghusa, John Wilson, David Norris, Tim O'Connell, Oscar Merne; Forest and Wildlife Service (now Dúchas), Dublin, Bray & Galway; Brendan O'Ciobhan; Ordnance Survey, Dublin; Diarmuid O'Laoghaire SJ, Milltown Institute, Dublin; Bros. Colman and Anthony, Glenstal Abbey, Co. Limerick; Frank Lyon, American Museum of Natural History, New York; Fr Cathaldus Gibney, Franciscan Library, Killiney; Christophe Darcet, French Embassy, Dublin.

Particular thanks to the following who helped with the text or provided illustrative material: C.D. Deane OBE, Belfast; Gerard Quinn, Blackrock; Fergus Kelly, Dublin; John Love, Inverness; D.J. Hall, Cambridge; Pierre Petit, Bordeaux; André Chabaud, Paris; Padrig an Habask, Brittany; Jim Fox, Kildare; Cilian Roden, Galway. Also, Frank Mitchell, Very Revd D. Woodworth, Erin Gibbons, Lord Revelstoke, Dale Sergeantson, Patricia Curran-Mulligan, John Temple-Lang, Noeleen O'Connor, Niall Murphy, Paul Burke-Kennedy, Paul and Donal Murphy, Jim Haine, William Gaynor; the Irish Raptor Study Group, Grúpa Staidéir Éan Creiche na hÉireann; Peter Vincent, Lancaster University; Duncan McDonald, Lorcan O'Toole, Perthshire, Scotland (particular thanks to Lorcan O'Toole for taking time out to introduce Tim Danagher, Paul Reynolds and me to the wonderful raptors of Perthshire during a long weekend in April 1999).

I would also like to thank the staff of the following establishments and institutions who, in one way or another helped with my research: Linenhall Library; Public Record Office; Queen's University, Belfast; Ulster Museum, Belfast; Royal Dublin Society; Society of Antiquaries of Ireland; Public Record Office, Northern Ireland; Land Commission; Royal Irish Academy; University College Dublin Library; Marsh's Library; Trinity College Library, Dublin; University College, Galway; G.P.A. Bolton Library, Cashel; University Library, Cambridge; Science Museum Library, London; British Library; Centre for Extra-mural Studies, Birkbeck College, University of London; RSPB, Scotland; Kirkwall Library, Orkney; Natural History Museum, Eua Island, Tonga; Louvre Gallery, Paris; Natural History Museum, Reykjavik, Iceland; American Museum of Natural History, New York; and the Field Museum of Natural History, Chicago.

Finally I would like to thank my wife Esther-Mary for her support, patience and the soundness of her advice, Theresa O'Brien who typed up the original text, and Professor Paul Nolan who had it transferred to disc.

LIST OF ABBREVIATIONS

ASI	Area of Scientific Interest	NIBA	Northern Ireland Birdwatcher's Association
BTO	British Trust for Ornithology	NPWS	National Parks and Wildlife Service
BWI	BirdWatch Ireland	OS	Ordnance Survey
EC	European Community	REPS	Rural Environment Protection Scheme
EU	European Union	RIA	Royal Irish Academy
HMSO	Her Majesty's Stationery Office	RSPB	Royal Society for the Protection of Birds
IPCC	Irish Peatland Conservation Council	SAC	Special Area of Conservation
IRSG	Irish Raptor Study Group	SPA	Special Protection Area
NARGC	National Association of Regional Game Councils	TCD	Trinity College Dublin
NHA	National Heritage Area		

INTRODUCTION

A deficit in the bird tally of a country is, regrettably, not unusual, at least not in the developed world. Most European countries are without a few of their former birds, though they are usually to be found elsewhere. Some regions, however, notably North America, have lost birds – Heath Hen, Labrador Duck, Carolina Parakeet, Passenger Pigeon etc. – which are sadly also lost to the world. Ireland, as far as we know, has only one extinct bird – the Great Auk – and fate has spared us the ignominy of hosting its extinction.

Birds are organisms, and organisms are constantly being lost, are they not? Species disappear from the world all the time. Why should we be especially concerned about birds? Why differentiate between a handful of birds and, say, invertebrates: where are the books about the preservation of mosquitoes? Or rodents? Surely they too have a place, a role in the overall scheme of things.

We might be more open to this logic if we could detach ourselves from our subject as science says we must. But just as we have always had reservations about science, we have never been even-handed about nature. 'A feathered, warm-blooded vertebrate which lays eggs' is as inadequate a description of a bird as 'a stringed musical instrument made of wood' is of a violin. The bird has consistently amounted to more than its component parts and in human consciousness has been the subject of long-standing admiration and mystery. Aristotle's fascination with birds more than two thousand years ago was merely an update of a relationship as old as sentient man. How many words have been written about birds or derived from their presence, how much music has derived from their songs, how many pictures, photographs, sculptures have they inspired? Birds have profound symbolic potency: they are arguably the most universally celebrated form of nature.

The disappearance of a bird, therefore, from a neighbourhood, a region, an island is not simply a reduction in biodiversity, a further ecological impoverishment, or even a depletion in the 'otherness' of nature; it is a cultural loss. A strand of the luminous fibre which interconnects all life, and radiates around us, is extinguished with the disappearance of a familiar bird. We ourselves are the duller for its passing.

When I began researching this book more than twenty years ago there were plenty of Corncrakes in Ireland. They were still widespread in Europe though they had declined significantly since the turn of the century. England had seen a drastic decline, and only a scattering remained on mainland Britain, mostly in western Scotland. Was this to be the pattern also in Ireland? At that time such a change would have been unthinkable, so quintessential an Irish bird was it: but time has not been kind to *An Traonach*.

Now, on the threshold of a new millennium, the Corncrake is almost gone from Ireland. If, despite the heroic counter-efforts presently under way, it does disappear completely, that will mark the end of an extraordinary socio-environmental record, which has its beginnings in prehistoric times (there are references to the Corncrake in the ancient glossaries and who knows how many centuries of oral history prior to that).

We know from archaeological evidence – bones recovered from ancient refuse heaps (middens) – that the Corncrake has been a source of food for people throughout prehistoric times. Its reluctance to take to the air rendered it easily caught in nets. Its secretive ways were undoubtedly the subject of discussion for Mesolithic hunters much as they are for present-day conservationists. The Corncrake has thus an illustrious pedigree. It has been in Ireland at least as long as we have and probably much longer.

Its *curriculum vitae* reads as follows: an item of food (up until the present century), a literary symbol (satire in ancient writings and sadness in recent poetry), a part of folklore, an inspiration for artistic expression, from music to painting, and since its decline the subject of concentrated scientific research. The Corncrake still remains, but it is now no more than a vestigial element of our heritage.

This book is about eleven other birds which are no longer part of our natural heritage. It is debatable whether any of them was as symbolically interwoven into the Irish psyche as the Corncrake. However, they were well enough known to have had specific Gaelic names and to have left echoes in literature and art and, in most cases, archaeozoological evidence in the ground.

It is probably too early to interpret the significance of these losses, for depletion tends to be assessed by using the hard-nosed monitor of statistics: they represent less than 10% of our total of 'common' birds. It is important, however, to look at them in collective context. For example, the six birds of prey that have disappeared (two eagles, Goshawk, Red Kite, Marsh Harrier, Osprey) represent half of the total which the country once supported at its optimum – and the larger, more spectacular, half at that. In terms of the effect on the food chain the ramifications are compelling: what would be the balance of our birdlife if these raptors were still around; does this help to explain why we have so many 'pest' birds like Hooded Crows?

The effects of geographical isolation and the subtle persuasions of climatic change have undoubtedly had a bearing on the decline of some of our birds, as they have on island communities throughout the world. These often unquantifiable factors have been well documented elsewhere, and some of the research is clearly relevant here. But other factors which recur in the case studies have been of more immediate significance. Most important has been direct and indirect human pressure, especially hunting and habitat change. The advance of technology in the 19th century – the arrival of breech-loading guns, sophisticated traps and poisons – and the consolidation of a game-orientated demesne system radically changed the nature and effectiveness of hunting. Most of our birds of prey which came into direct conflict fell foul of this advance. The drainage of formerly extensive wetlands (particularly from the mid 19th century onwards) and the clearances of the last remaining tracts of native forest have been undeniably important. In this latter instance the developmental surge of the 17th and 18th centuries appears to have been crucial. In each of the case studies consideration has been given to the relative importance of habitat destruction and modification.

The other side of the coin shows that Ireland has welcomed 'new' birds down the ages. Much of this re-enrichment has been inadvertent, the consequence of introduction by man. The Pheasant and the Mute Swan, members of our fauna since the Middle Ages (along with the Rabbit and the Fallow Deer), were brought here for the chase and the pot. Others, like the American Turkey, introduced here in the 17th century, have not survived in the wild. A number of birds, of their own volition, have simply extended their range to include Ireland, found it to their liking and stayed. Most familiar is the ubiquitous Magpie, whose arrival here, in Wexford, in the latter 17th century is recorded in the literature. The Fulmar and the Eider Duck moved here from further north about a hundred years ago, and the Collared Dove, although it has the appearance of a long-term resident, arrived in Ireland from the south (originally Turkey) as recently as 1959. The Little Egret, which has established a colony in Cork in the closing years of the millennium, is a spectacularly attractive newcomer. Celebrated and enriching as these newcomers are, however, they cannot be viewed as substitutes for our losses.

We must also celebrate a bird that came back. The Buzzard, like the Red Kite, soared over the prey-rich woodlands until, having been stripped of its habitat, was hunted to extinction with poison and trap in the 1890s. The development of a more accommodating attitude towards predators in Britain during the war years resulted in the Buzzard returning again to Ireland. Successful breeding in Co. Antrim in the late 1950s was followed by spread to many parts of Ulster. Nowadays this magnificent bird of prey shows signs of

1 *Buzzard: body and flight*

recolonising the Island once again. But Buzzards (and kites) are still being poisoned by ill-informed individuals, and if we are to have this raptor as it once was and fulfilling a much-needed role in our countryside, it will take a degree of awareness, tolerance and far-sightedness beyond that which is presently evident.

The return of the Buzzard may be seen as an example of what is possible if only the right circumstances are provided. We should make it an aim to recreate the circumstances whereby most, if not all those listed (excepting, of course, the Great Auk) can be reinstated here. Our economic connection to Europe has seen the development of a more enlightened land-use policy, with serious consideration at last being afforded to conservation. There has probably not been a better time to put in train a coordinated reintroduction programme. Due to lingering attitudinal problems on the ground, particularly in regard to birds of prey, it might be decades before such a programme is widely accepted. Nevertheless it is long overdue. The attendant environmental and cultural re-enrichment would undoubtedly benefit us more than we can presently imagine.

I hope *Ireland's Lost Birds* will contribute to this new beginning.

PART ONE

BACKGROUND

In re-establishing the story of our lost birds it was necessary to delve extensively. Much of the modern source material lay near the surface (where it would be expected) in the natural history literature. It was clear, however, from the brief, sometimes contradictory, references by ornithologists that it would be necessary to look outside the normally consulted sources.

Social historical literature revealed frequent passing references to formerly familiar birds and, in some instances, data relating to the chronology of their disappearance. These extracts provided much of the basis for this book, but artefactual material also helped in the reconstruction process. While some of this was, of itself, revelatory, much simply added weight to the literary evidence.

A good deal of sound natural historical material had already been collated by naturalists and was accessible in the pages of ornithological and zoological journals. There are chapters and extracts in the natural history texts on the subject of extinct Irish birds. Most of these emanate from the Victorian era when forefathers of the natural history movement became curious or concerned about our changing avifauna. While this corpus of nature literature is reliable back to the turn of the 18th and 19th centuries, one gradually loses confidence in it as we delve into earlier work. Not that it should be dismissed altogether, for there are well-substantiated snippets from as far back as the 12th century and even earlier.

Three medieval images of Irish Cranes

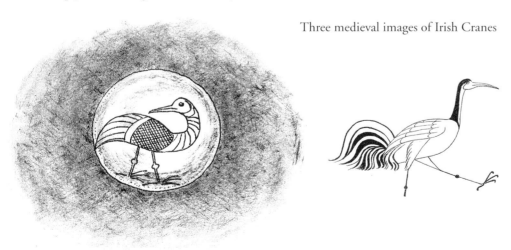

2 *Viking coin, Wood Quay, 10th–11th century*   **3** *Drawing from* Topographia Hiberniae, c.1200   **4** *Tile from Mellifont Abbey, 14th–15th century*

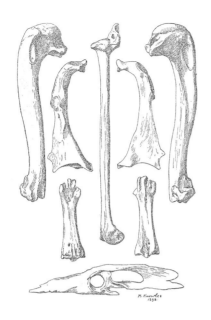

5 *Bones of the Great Auk*

As some of the lost birds evidently disappeared before the year 1800 it was necessary to look critically at Early Modern nature references. Fragments of useful information, in English, are contained in manuscripts and sundry other documents, particularly since the time of the Tudor Plantation. The various surveys, tours and topographical descriptions of Ireland contain evidence, as do statutes, patent rolls, early legal documents and estate letters.

The wealth of Irish literature in the form of ancient poetry, prose, extracts from the annals and the folklore yielded much material. As these sources are, thankfully, mostly available in English, the task was made infinitely easier for someone with little Irish. It was, however, necessary to access the original Irish in some cases, to be sure that the translation was correct, and to those scholars (listed in the acknowledgments) who helped me I again express my gratitude. Some contributory information emerged from the placenames and townlands with ornithological connections,

which, although mainly corrupted into an English form, could nevertheless be easily reconstructed into the Irish for deciphering. Early Modern Irish (pre–1650) was particularly important as it is apparently largely standardised. Modern Irish with its regional dialects and variations (though undoubtedly interesting and often enriching) often had the affect of complicating instead of clarifying.

Artefacts in the form of illustrations in some of the historical documents were useful, but due to stylising and distortion it was necessary to exercise caution (as well as imagination) in their interpretation. Carved functional or decorative pieces and heraldic symbols from the the Middle Ages provided some useful background material.

Bird bones uncovered in archaeological digs provided most of the zoological background, a good deal of which pre-dated the literary evidence. I am greatly indebted to the work of a new genre of archaeologists who, having collected faunal bones from archaeological sites, made species identification a priority. Bones may be described in the text as 'fossils' and the bone evidence as 'fossil evidence'. They (even the oldest) are, of course, not fossils since they are invariably original remains and not the traces formed by complete mineral replacement. Some journals refer to them as 'subfossils' but I find myself avoiding that term. I hope I will be forgiven this liberty, taken in the interest of readability.

Bird bones were uncovered, often in large numbers, from sites of man's early habitations. Others were found in places such as caves where they were possibly the meal remains of predatory animals. Many of the early sites were kitchen middens (the eating places of hunter/gatherer groups). These ranged across the prehistoric/historic divide but were generally prehistoric. Others – crannogs and early Christian dwelling places – provided data from the first millennium AD. Tower house bawns and early town dumps were productive in certain instances, particularly for bones of scavenging predators. Archaeozoology is still an infant science and there is obviously a great deal to be learned of our ancient flora and fauna through its development. Doubtless future excavations will reveal a great deal more, some of which cause us to reconsider our present ideas about our past wildlife and our predecessors' relationship with it.

Most of the fossil evidence corroborates early literary evidence but some raises further, so far unresolved, questions.

Through the examination of faunal evidence from locations such as Mount Sandel (near Coleraine), Wood Quay, Newgrange etc. a picture is emerging of how our predecessors lived, what they ate and what wild and domestic animals and birds cohabited this island with them.

Peat bogs, repositories of so much preserved knowledge, have been treasure chests. Palynological studies (pollen analysis) from bog cores have helped enormously with the reconstruction (at least in the imagination) of previous habitats. It has been especially useful in the investigation of former woodland cover – an important factor for several of our lost birds.

RELIABILITY OF BACKGROUND MATERIAL

While the question of literary reliability has undoubtedly deterred serious investigation of this kind, it is not an issue with the interpretation of the fossil evidence. Assuming the expertise of the archaeologist involved, it is entirely reliable. Birds can be identified from a few, or in some cases, even a single bone. Identification is carried out by comparison with a skeleton. It would be necessary to have access to some hundreds of bird skeletons for this purpose. Unfortunately such a collection does not exist (to date) in Ireland, and identification is usually established by sending the bones to Britain or elsewhere where such collections exist for comparative purposes. *Bird bones from archaeological sites* (Cohen and

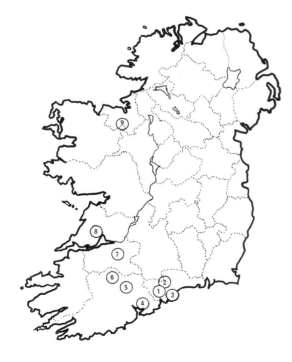

1. Shandon
2. Ballynamintra
3. Kilcreaney
4. Ballyfin
5. Castletownroche
6. Castlepook
7. Lough Gur
8. Edenvale, Newhall
9. Keshcorran

6 *Cave sites which produced bird bones*

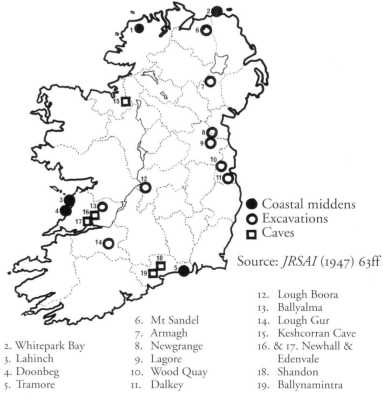

● Coastal middens
○ Excavations
□ Caves

Source: *JRSAI* (1947) 63ff

2. Whitepark Bay
3. Lahinch
4. Doonbeg
5. Tramore
6. Mt Sandel
7. Armagh
8. Newgrange
9. Lagore
10. Wood Quay
11. Dalkey
12. Lough Boora
13. Ballyalma
14. Lough Gur
15. Keshcorran Cave
16. & 17. Newhall & Edenvale
18. Shandon
19. Ballynamintra

7 *Archaeological sites with abundant bird bones*

Serjeantson, 1996) enables reliable identification to be carried out by non-experts. Undoubtedly literature such as this will, over time, add substantially to our increasing store of knowledge.

Some of the early work carried out by A.W. Stelfox and his contemporaries in this field should surely be reassessed in the light of technological developments. For instance, identification of game bird bones from one or two caves has not been positively established, and bones, supposedly those of Ptarmigan and Black Grouse, have since been reassessed as having come from other, more likely, species. Even when identification is established beyond doubt a number of questions remain. Was the find an isolated case? Was the bird in question brought from elsewhere to the place of discovery? and so on. In Stelfox's day, dating of the fossil evidence was estimated by stratification (by its location in the ground). Nowadays it is carried out accurately but expensively by Carbon-dating. This process can, in most cases, establish the date of death of the bird to a margin of a few decades. One problem is that a reasonable quantity of carbonaceous material is required and it is completely destroyed in the process. Still, where large quantities of bones are concerned (as at most kitchen middens), the loss, in terms of percentage preserved, is minimal and usually does not prevent dating being carried out.

Artefacts and illustrations from the distant past also require cautious consideration, The stylising of these motifs, the possibility of importation from abroad and other factors make their interpretation far from straightforward in most cases.

The problem of reliability is most acute in the case of the literature, for the natural history of Ireland is at best incomplete and mostly non-existent further back than the past two centuries. We owe a dept of gratitude to the early 19th-century naturalists Templeton and Thompson who wrote meticulous and accurate accounts of contemporary flora and fauna and provided us with a solid ornithological bank to which we can return again and again. Though reliability may be called into question in more recent work, it is possible by falling back to those earlier naturalists to interpolate reasonably accurately.

Table 1 *Bird list from Wood Quay excavations, 10th–11th century*

Species	No. of fragments	MNI*	% of total MNI
Fowl	511	58	36.47
Goose (all species)	88	16	10.06
Raven	231	21	13.21
Buzzard	51	10	6.29
Crow	53	7	4.4
White-tailed Eagle	32	5	3.15
Oystercatcher	10	5	3.15
Herring Gull	14	5	3.15
Kite	10	4	2.52
Cormorant	7	3	1.89
Peregrine Falcon	5	2	1.26
Osprey	7	2	1.26
Swan	4	2	1.26
Hen Harrier	4	2	1.26
Marsh Harrier	4	2	1.26
Greater Black-backed Gull	5	1	0.63
Kittiwake	1	1	0.63
Jackdaw	2	1	0.63
Gannet	1	1	0.63
Curlew	2	1	0.63
Shag	1	1	0.63
Bar-tailed Godwit	1	1	0.63
Guillemot	2	1	0.63
Duck (all species)	5	3	1.89
Sparrow Hawk	1	1	0.63
Red-throated Diver	1	1	0.63
Rook	1	1	0.63
Crane	2	1	0.63
Assorted unidentified waders	20		
Assorted unidentified others	268		

* MNI = minimum number of individuals

Eighteenth-century Irish ornithology was documented to varying degrees of reliability by a series of naturalists. None commented as comprehensively as the English naturalist Pennant, but some including Smith, Harris and Rutty made valuable contributions despite the constraints of the time. Some of the individuals who wrote regional histories or who toured extensively in Ireland also recorded data on the birds they encountered. Young and his 18th-century contemporary Pococke did so. O'Flagherty in the later 17th century provided information in Irish but, as in most of the accounts of this period and earlier, it is supplemented with imaginative and fanciful comments. However, in view of the emphasis on myth and the great desire to explain the unobservable among the Irish (and most contemporary peoples), interpretation of a kind is not valueless. Thomas Molyneux, far from baulking at the propensity among the county people whom he questioned, recorded some valuable insights which are noted in his collection of papers on the natural history of Ireland (1684).

One work which stands out is the *Zoilomastix* of Philip O'Sullivan Beare (*c*.1625). It was written in Latin (probably in Spain) but has the names of most of the birds in Irish (in the margins). Despite the mythological embellishments and fanciful classical allusions, the author's obvious first-hand knowledge and his list of the Irish names makes his manuscript an important commentary on Ireland's birds from nearly four centuries ago.

A series of Tudor observers in Ireland wrote about or briefly mentioned the natural history including the birdlife in Ireland in the second half of the 16th century. These allusions were often comments on the wealth of natural resources which the native Irish tended 'foolishly' not to exploit for food. Spenser, who mentioned birds and habitats in his ornate poetry, was a spokesperson for this view, but we are never sure whether his comments relate to his Kilcolman estate or to his English homeland. Other commentators, including Derricke, Moryson and Boate, supplied fragments of a more prosaic kind. There are also a number of English references to Irish birds from Tudor and Jacobean writers who may never have set foot in Ireland.

Higden, in the 14th century, mentioned Irish birds but his information was mainly from earlier references, particularly those of Giraldus Cambrensis. A great deal of ridicule and criticism has been levelled at Giraldus, a Welsh monk who came to Ireland in the 12th century, and his work *Topographia Hiberniae*, which is rich in bird references. Most of the criticism is due to the imaginings and fables with which his books are liberally sprinkled. He was, of course, simply continuing a tradition (begun by Aristotle and his contemporaries) to explain the inexplicable in the centuries before the scientific enlightenment. It is important in attempting to interpret Giraldus' work to try to put oneself into his shoes; after all, his mission as chronicler for Henry II was to record what he saw, what he thought he saw or what he was told. A hotchpotch of fact and fancy was inevitable. It is nevertheless possible to extract circumstantial material which can often be substantially corroborated elsewhere. Despite its shortcomings, the section on birds remains the earliest account of the present millennium that exists today.

Still further back is the early Christian monastic literature, the early (mainly pre-Christian) mythology and the Brehon laws. The early Christian monks were fascinated by nature, and their poetry is significantly drawn from the flora and fauna of the time. Difficulty exists in the translation of Early Irish, and some of the abundant commentary about the fauna of the 8th to the 12th centuries is obscure and not open to specific interpretation, The ancient Irish laws, particularly *Breatha Comaithchesa* (laws of neighbourhood) and *Breatha Forma,* contain allusions to birds. The translation and actual identification of the species intended remains a problem, and careful interpretation is required in nearly all of this early work. Fergus Kelly's painstaking exploration of the laws has been both timely and invaluable (*Early Irish farming*, 1997).

Understandably, commentary on Irish birdlife is non-existent prior to the early Christian period. We have to rely on isolated artefacts (such as the wonderful La Tène-style Wren found at Dun Aengus) to carry the investigation into prehistoric times. There is other bird-motif material from as far back as the Bronze Age but it adds little, overall. To all intents and purposes our cultural reference is an historic one pertaining mainly to the past 1,500 years.

It is to be hoped that most readers will accept the evidence as presented, taking account of the limitations under which it was written. Some may take a different perspective, based perhaps on the incompleteness of the evidence in some cases, and draw conclusions other than those to which I have come. Whichever the approach, this motley collection of material constitutes the record of our birdlife from the earliest times and if we are to piece together the relevant natural histories and investigate the factors which brought about losses, all we can do is tip the bucket and sift through the contents as they are presently available to us.

NOTES

Bird names in the Irish language

The following are the main sources of bird names.

Early and Middle Irish (9th–11th century)
Translations of poems, legendary tales, triads, Brehon law texts, glossaries (Cormac, Egerton, O'Davoren).

Modern Irish
Dindshenchas poems.

17th-century
(*c*.1625) P. O'Sullivan Beare, *Zoilomastix*
(1662) R. Plunkett, *Latin/Irish dictionary*
(1685–90) W. Bedel, Translations of the Bible

18th-century
(1739) J. K'Eogh, *Zoologica Medicalis Hiberniae*
(1739) T. O'Nachten, *Irish/English dictionary*

19th-century
(1828) The Diary of Humphrey O'Sullivan
(1842) A. Mhic Dhomnaill, *Fealsunacht*
(1849–51) W. Thompson, *Natural history of Ireland*

20th-century
(1900) Ussher & Warren, *Birds of Ireland*
(1901) C.W. Benson, *Our Irish songbirds*
(1915) R.J. Scharff, 'On the Irish names of Birds' (in *J.N.S.*)
(*c*.1940) (unpubl.) A. Ó hAonghusa, 'Collection of Irish names of birds'
(1967) C. Moriarty, *A guide to Irish birds*

FACTORS INFLUENCING LOSSES

Extinction is a natural phenomenon. 'Background' extinction accounts for several species of higher animals, including birds, every million or so years. Of course, this depletion is counterbalanced by the evolution of new species over a similar vast timescale. Natural history reflects an overall tendency towards equilibrium.

Nonetheless there have been distinct episodes of 'catastrophic' extinctions. Five are often cited – extending from the Ordovician (440 million years ago) to the well-publicised Cretaceous episode which did for the dinosaurs 65 million years ago. Presumably there were many others which we now know nothing about. The causes of these devastating events remain obscure: there are probably as many advocates for extra-terrestrial activity (asteroids etc.) as there are for sudden, random climatic change. Given the amount of investigation and comment that the controversy has spawned, the eventual emergence of a consensus seems improbable.

Though birds have been around since the age of the reptiles, 160 or so million years ago, it was not until the early Tertiary era, some 40 million years ago, that the first major groups – diving and wading birds – became established. Mass extinction has been traced back almost to this time: palaeontology has revealed traces of a suite of tropical birds which, in response to the changing climate of the early Oligocene, vanished from Europe, never to return. Our perching birds, insect and seed-eating types like warblers and finches emerged later – perhaps 10 to 20 million years ago. The fossil record has revealed that the major bird orders typical of Europe today (beginning with divers and grebes: ending with finches and sparrows) were present prior to the beginning of the last Ice Age, perhaps two million years ago. There are grounds

to believe that subsequent extinctions, while they undoubtedly occurred, were the exception rather than the rule, despite the wide climatic fluctuations of the period.

We are still in the dark as regards the avifaunal changes which must have occurred following the climatic ameliorations of the Postglacial period and the early Bronze Age, when average temperatures were perhaps one or two degrees warmer than today. The temperature drop of the 'little' Ice Age (a centuries-long climatic 'blip' which peaked around 1750 AD) must also have had now-unrecorded effects, particularly in Britain and Ireland.

Burton (1995) identifies a change in Britain's bird population, due to recent climatic variation. Northern species – Red-necked Phalarope, Dotterel, Snow Bunting etc. – are gradually retreating northwards; southern species – Serin, Cetti's Warbler, Collared Dove – are advancing. The issue is confused, claims Burton, by certain southern birds – Wryneck, Cirl Bunting, Stone Curlew – which, due to recent wetter summers in the south of England, are retreating to the south. It is difficult to see parallels in Ireland due to our lesser latitude range and our consistently wet summers, but the recent colonisation by the Little Egret on our south coast may well be associated with the south-to-north trend.

What are we to make of the predictions for the future influence of climate? We are told that global warming will determine the shape of our avifauna, and if we accept Burton's thesis this may be happening already. Other observers are preparing us for the return of the ice. Evidence has been mustered which purports to indicate a steady cooling since the mid 20th century. While the question remains unresolved, most scientists accept that man-induced warming, at any rate, is a reality. Our lack of consensus seems to stem from the fact that the parameters exhibit a reluctance to stand still long enough to be quantified.

Our image of how climate change will affect our birdlife must take account of analytical shortcomings. Theories and mathematical models are inclined to regularise climatic considerations, to view them too locally and to underestimate how inextricably they are interwoven with geography.

Geographic circumstances have an important bearing on extinctions. They are much commoner on isolated islands than elsewhere. Ninety per cent of all recent bird extinctions have occurred on islands. Hawaii, for instance, has lost 24 birds. The island theory of biogeography as espoused by McArthur and Wilson and others has helped with our understanding of the processes of invasion and extinction, but distinction must be drawn between island 'isolates', like Hawaii, and 'continental' islands like Ireland.

Geographic separation, as distinct from isolation, is highly pertinent in Ireland, where our fauna occupies an island off an island off a continent. We have been reliant on the stepping-stone of Britain for a substantial fraction of our land-based wildlife which would otherwise not have made it here. Frank Mitchell's 'steeplechase' model demonstrates that many species of flora and fauna (including non-migratory birds) inhabit Ireland only because they were able to clear a series of natural obstacles – sea crossings and mountain ranges.

Ireland is not remote in ornithological terms: we do not have endemic species (only sub-species or races: Red Grouse, Jay, Dipper, Coal Tit) like so many of the Pacific islands. Most of our large seabird colonies have established themselves due to our ocean-fronting western seaboard, but our woodland passerines, many of which have sedentary lifestyles, must have had difficulty spreading to Ireland and must have come from the south and east. Many birds thus inhabit Ireland at the edge of their range. It is known that in this circumstance they tend to occupy, as a preference, the best available breeding sites, whereas in the centre of their range they will occupy even marginal ones. This has made them more vulnerable to extinction when subject to extraneous pressures.

As a smaller island Ireland has probably always had an impoverished avifauna compared with Britain. Island biogeography states as much. Populations of birds with specific habitat preferences must therefore have been low, since smaller islands tend to have less habitat variety. A critical 'low' is cited for such island populations. When there is a drop below a certain number (between 50 and 500 depending on species criteria) in the absence of extraneous support, a genetically waning population tends to teeter towards extinction. There are examples from islands throughout the world: New Zealand's Takahe comes to mind. However, as the Takahe has also demonstrated, even flightless species can continue to survive in circumstances of seclusion. Nor should we forget the Mauritius kestrel, at one time the rarest bird in the world. According to Minimum Viable Population theory, the kestrel had declined far beyond the level of recovery, having dwindled to less than 10 birds in the wild. Dedicated intervention of an ecologist, however, saved it from extinction and now, with more than 100 pairs on the island, its future looks secure.

Extinction is often complex: its mechanisms tend to defy theoretical considerations.

In most cases extinction has been accelerated by the inadvertent or deliberate introduction of predators, and exploitation by man has added an extra, usually lethal, dimension.

It is not the brief of this book to weigh the pressures on island bird groups or to apportion blame. Suffice to say that, in the history of extinctions, world-wide climatic, geographic and other 'natural' factors are usually thought of as being secondary to the 'human' factor. In most cases extinction is thought to have occurred due to a combination of both. It is pressure brought about directly or indirectly by humans that is most often pondered and, sometimes, simplistically isolated. In the discussion in this book humankind may be presented as nature's unconcerned and irresponsible exploiter. While there is no denying the ubiquity and longevity of the notion of nature as the inexhaustible resource, humankind has also identified with it culturally and sometimes sympathetically. There is evidence for this in Ireland, as elsewhere in the inhabited world.

As far as we know, people did not reach Ireland until well after the end of the Ice Age – very late by the standards of other European countries, including Britain. Earliest traces here date from the Mesolithic (Middle Stone Age) about nine thousand years ago. These hunter/gatherers lived off whatever birds and animals they could catch, supplementing their diet with shellfish and edible plants. Their dining places have been found around the coast, and the debris of their 'picnic' middens have been found, mainly on the coast but also far inland. Bone remains at these sites have shown that these early men caught and ate birds of a range of species, including Great Auks.

The first people to farm land here began about six thousand years ago and there is much speculation as to the rate at which this activity progressed in view of the fact that Ireland was largely an impenetrable wilderness of mature woodland and wetland at the time. Tree clearance (mainly of pines) was carried out and crops were planted. Archaeological investigation has revealed that this early farming originally took the form of 'shifting agriculture' – small areas of forest being cleared and the process repeated after the fertility of the land had been used up. The resulting natural regeneration of the tree cover and landscape change would have been so gradual as to have had only a local, temporary affect on the flora and fauna. The network of stone walls beneath the blanket bog in the west of Ireland indicates that some extensive clearance was undertaken by these settled (mainly livestock) farming communities. This, combined with a change to cooler wetter climatic conditions, may have facilitated the growth of blanket bog, five thousand years ago.

The gradual clearance of the pine forests which for three thousand years dominated both upland and lowland, probably permitted hardwoods to become dominant on the thicker soils of the lowlands. The pollen record shows that one species, Elm, suffered a devastating decline towards the end of the Neolithic: whether caused by Dutch elm disease (as suggested by Mitchell and others) or not , the disappearance of the elm, a major element in the climax tree cover, enabled inroads to be made into formerly impenetrable hardwood forests.

The well-documented population growth of the later Bronze Age, combined with the arrival of the ard plough, enabled considerable agricultural development on newly cleared land. The location of the settlement sites and their megaliths – generally near the coast, on uplands, or along the more accessible river valleys and eskers – indicate that man's influence continued to be somewhat peripheral and that much wilderness remained in Ireland throughout prehistory. It was in regions like the Burren, where a thin tree covering and well drained soils permitted easy access for farming, that population was concentrated.

A period of population decline occurred in the Iron Age and regenerating woodland returned, obliterating much of the cleared land. For several centuries Ireland's landscape appears to have reverted to that of the Neolithic.

Woodland mammals such as Wolf, Wild Boar and Red Deer, Pine Marten and Red Squirrel and large birds such as the Capercaillie and woodland raptors continued to thrive. With the possible exception of the boar, all are known to have survived well into historic times. Sufficient wildwood (extensive mature forests) must have survived, therefore, to have provided sanctuary to the undepleted post-glacial avifauna.

WOODLAND CLEARANCE

From Early Christian times onward Irish society began to impact permanently on the landscape. While not directly affected by Roman technology and organisational skills, the rural Gaelic regime was nevertheless experiencing island-wide development. The establishment of the ringfort farm system – some 50,000 were built throughout the country between 500 and 1000 AD – transformed much of the landscape. Swathes of the regenerated, secondary woodland were removed to allow for the creeping advance of the farmsteads.

The momentum of the Early Medieval does not appear to have been sustained into the latter centuries of the period. It is clear that the population of this country increased at a slower rate than in neighbouring Britain. This was partly due to a massive loss of life in the 14th century due to bubonic plague, from which the medieval population never really recovered.

The population of Ireland at the decline of the Gaelic order in the early 17th century was not much in excess of one million. The slow rate of development may also have been due to the attitude of the Irish towards natural resources which appears to have been driven more by local requirement than by external market forces. The Brehon laws emphasise the judicious, selective use of timber. Certain tree species had specific uses for utensils, vehicles, house construction etc. and even the collection of kindling was itemised. Oakwoods were maintained for feeding pigs (on the acorn crop). Early Gaelic sources suggest that, besides the purely utilitarian, there existed a respectful social attitude towards woodland. This

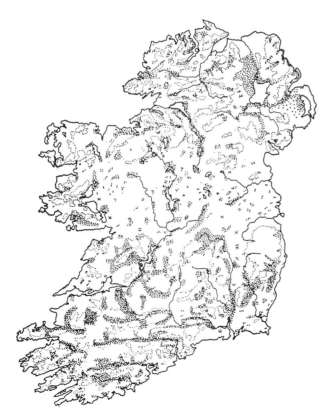

8 *The extent of woodland c.1600*

arboreal consciousness, clearly a legacy from early Celtic times, has been referred to by Neeson (1991) and Foster (1997).

The surviving poetry of the early Christian monks is rich in the appreciation of nature. So numerous are the names and descriptions that an awareness of the inspirational effect of the natural world on these people comes through – reflected in the illuminations of the manuscripts of the time. Not only did the monks interpret their environment but they protected parts of it as well. Could they be described as Ireland's first conservationists? The network of monasteries and other ecclesiastical houses throughout the country acted as protectorates for woodlands and other wildlife habitats until the dissolution of the monasteries in the mid 16th century. The fact that over three hundred major ecclesiastical institutions were in existence in Ireland until the 16th-century dissolution suggests that their influence was widespread and effective.

The Normans also brought with them a kind of respect for the land. Their vocabulary specified the characteristics of forest, chase, park, covert, warren etc. 'Forest', a term meaning a royal preserve, probably had its equivalent in the Gaelic system, though to the Normans it meant not only the treed region but also the treeless, open areas in the preserve.

The other terms relate to hunting and refer to specific areas set aside for the hunting of various kinds of game. During the early Middle Ages the Normans introduced the concepts of land management and organised husbandry and, like the monks, they acted to some degree as self-interested conservationists. It would be an unfortunate generalisation, however, to suggest that our environment was being positively protected by men of high ideals. The Gaelic Irish and the Norman settlers were often involved in wanton property and habitat destruction in the many land-grabbing and other fracas that were a feature of medieval life both here and abroad. The annals mention the cutting and burning of trees in attacks and reprisals. Nevertheless, it is hard to believe that species diversity was seriously threatened or significantly reduced at this time.

With the arrival of the Tudor settlers from the later 16th century onwards, sweeping permanent changes were wrought on the Irish landscape which had undoubted effects on wildlife richness and diversity. The pollen record indicates clearly that the most radical diminution of woodland cover occurred at around this time. As late as the beginning of the 17th century extensive hardwood forests (mainly of oak and ash) and scrubland were still to be found here: 'In 1600 about one eighth of Ireland was forested; by 1800 the proportion had been reduced to a fiftieth as a result of commercial exploitation of the Irish woodlands' (E. McCracken, 1971).

The remaining woodlands were systematically cleared for fuel for the new industries – iron-smelting, glass-making, house- and ship-building, and coopering – and for the removal of wolves. The new regime was utterly intolerant of wolves and set about their eradication by destroying their woodland retreats. By the time the mid 17th-century maps of the country were drawn up by Petty in the Down Survey, it was obvious that much of the forest was already gone. Between one and two million acres of mature woodland and scrub were cleared in little over a century, and the land made available was put into other agricultural use. The establishment of demesnes and estates and the passing of the Enclosure Act immediately after the Restoration ensured that the change was permanent. Pockets of woodland did survive within the estates, but the first Ordnance Survey revealed a country even less afforested than that of today, though with pockets of natural or semi-natural woodland still evident on the higher ground of Leinster and Munster and in isolated parts of Connaught. Ulster, having been the most wooded province in 1600, had hardly a stand of trees to speak of a hundred years later.

Plantations of introduced hardwoods can be traced back as far as the early 17th century when Sycamore was first planted and the latter 17th century when Beech was first planted. Scots Pine, having disappeared in Gaelic times, was replanted from the late 17th century on. Since that period an ever-increasing range of non-natives or exotics have been brought, and the native hardwoods have been replaced in most places by spruces and pines, natives of other countries like Scandinavia or North America.

Nowadays Sitka Spruce and Lodgepole Pine plantations are a feature of every province and county, and, although they attract and provide habitats for a few bird species, they are much less important to flora and fauna than hardwoods. Environmentally they are no substitute for the original woodlands.

In any inquiry into the former existence and range of now extinct woodland birds, the reconstruction of and gradual diminution of their habitat is important. The Capercaillie is a case in point. If, as is suspected, some native pine woods remained in more remote areas of Ireland until the final clearances, it is easier to comprehend how such a specialised bird survived as long as it did. This is not to dispel the theory that, with the disappearance of the native pines, this large grouse had not gradually adapted to life in the hardwoods which were a feature of the landscape until much more recently.

It has been shown theoretically that habitat depletion does not cause extinctions of birds (or other fauna) until the great majority of the habitat is gone. In other words, birds have the capacity to survive, to hold on, until the basic fabric of their requirements is in a state of collapse. Our native woodland, or wildwood, has been subject to clearance since Neolithic times. Over most of the pre-historic period, clearance was fragmented, and sporadic and woodland regrew on previously cleared ground. Pollen analysis has revealed that this was especially true over the centuries immediately before and after the time of Christ. It was not until the 17th century that clearance became widespread and permanent. It is no coincidence therefore that the majority of the woodland extinctions occurred before the end of the 18th century, by which time the woodland habitat had been reduced to a level beyond the critical.

WETLAND DRAINAGE

The other major habitat that was to be drastically altered was wetland. The lakes, rivers, marshes, fens, bogs and, in coastal areas, estuaries, saltmarshes and lagoons were so vast and untamed that the idea of controlling them, let alone modifying them for use, would have been laughable until a couple of centuries ago. It is difficult nowadays to envisage how much greater they were in extent compared to the present. Some of the Tudor and Jacobean survey maps and traveller accounts hint at the impassability and unpredictability of wetland 'morasses' particularly in winter. The controlling and eventual drainage of the wetlands represented an even greater challenge to the 17th-century colonists than did the woodland clearances. There are references to the attempted drainage of turloughs from as long ago as the mid 17th century, but, due to the vast areas of flooding involved and the lack of drainage machinery, these were no more than gestures. The great estates of the 17th century were established in areas of good land, the bogs and the marshes being ignored as mainly 'unprofitable' – a word often used in the Down Survey. At this time the Midlands and particularly the basin of the Shannon and its tributaries were dangerous, impassable wetlands for most of the year. A Drainage Act was passed in 1715 by the Irish parliament to try to reduce the area of winter flooding and to improve navigation, but it was only partially effective on the latter score. In the mid 18th century grants were offered by the Royal Dublin Society to encourage this kind of work, but despite the production of guidelines there can be little doubt that the task overawed most landowners.

The Commission of Public Works heralded a new confidence in the field of drainage in the early 19th century which was consolidated by the successful Shannon Embankment Scheme. The newly established Bog Commissioners drew up elaborate plans for the drainage of the Midland raised bogs also in the early 19th century. The culmination of these seminal efforts was the 1842 Drainage Act. Initially these works were highly labour intensive (tens of thousands of men received gainful employment) and were included in the famine relief scheme of the mid 19th century. With experience gained from abroad (mainly south-east England and Holland) the drainage works gradually became more mechanised and efficient, and by 1900 almost a million acres (400,000 ha) had been effectively drained under about 220 schemes. Accelerating agricultural development (a response to the ravages of the famine) ensured that field drainage and the comprehensive utilisation of the reclaimed land ensued. This mass drainage had obviously a very dramatic effect on the avifauna of the wetlands. Both the resultant drainage and the effect of years of summer disturbance at the breeding sites of the aquatic birds must have proved an intolerable pressure to sensitive species such as the Bittern and Marsh Harrier.

The 20th century has seen radical changes in agricultural and rural practices which have further pressurised our avifauna. The widespread exploitation of the bogs has, within the past half century, radically changed the formerly unmodified landscape of the midlands. Many species of flora and fauna have been reduced by this activity (an unfortunate side-effect of a highly successful and beneficial industry). The accelerating activities in the Midlands over the past few decades have given much cause for concern. Had it not been for the work of the IPCC, which single-mindedly undertook to save intact a handful of representative examples, there would be have been no unworked raised bogs left in Ireland by the year 2000.

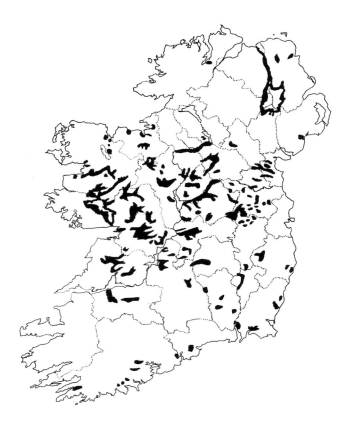

9 *The extent of drainage under the 1842 Act*

The 1945 Arterial Drainage Act saw a resurgence of interest in drainage works which concentrated in lowering the water table throughout entire catchments rather than along the watercourse itself. As a consequence of this new drainage a further one million acres (400,000 ha.) of wetland has been affected (some was land that due to neglect had reverted to wetland again). Besides the direct effects, thousands of acres of bog were provided with outfalls, rendering them available for exploitation. The benefit to the rural community in jobs and development seemed great in the initial stages. However, since the enthusiasm of the 1950s and 1960s it became increasingly evident that the benefits of drainage did not necessarily outweigh the costs and that the money spent on the schemes would, in many instances, have been better spent on improving already viable land. Due to the steady fall-off in crop-growing, particularly in the West, and the generally half-hearted 'follow-up' field drainage of the land, the whole question of arterial drainage in Ireland is under continuous review by our decision-makers, and the likelihood is that it will soon be wound up completely.

Despite the radical nature of the arterial drainage programme it is unlikely to have caused the extinction of any of our breeding wetland birds. It hit wildfowl hardest. Our wintering ducks, swans and geese lost many traditional haunts. Most adversely affected was the Greenland White-front, our 'bog goose', which, with the drainage of its marshes and the working of its bogs, was ousted from most of its traditional haunts.

AGRICULTURAL MODIFICATION

Changes in agricultural practice have also had serious and long-term detrimental effects on Ireland's birdlife. For as long as records have been kept, Ireland has had three native game birds, the Quail, Partridge and the Corncrake, which have cohabited with man on cultivated or semi-cultivated land. The 20th century has seen a steady decline in all three, and it is difficult now to see how they can avoid becoming unfortunate additions to the list of lost birds. The decline has been occurring simultaneously throughout western Europe and it would appear that modern agricultural methods are, at least partially, to blame. Other factors, like fluctuations in climate which have been affecting insect and plant requirements of these birds, especially at the fledgling stage, have been put forward as reasons for the decline, but opinions vary as to their relative importance.

The advent of modern agricultural methods has put increasing pressure on some more retiring species while favouring others. Silage cutting, now carried out two or three times in a season in many parts of the country, is ensuring that 'ground birds' cannot coexist easily with the farmer in these areas. On the other hand, the open field mechanised farming technique is favouring the Woodpigeon, which has reached pest proportions in many places. The general tendency towards larger farms and more mechanisation has meant the destruction of hedgerows – prime wildlife habitat. Indeed, this is proposed as a primary factor in the decline of Partridges in Britain and northern Europe. It has undoubtedly affected the populations of many plants and insects and has been agriculturally counter-productive in causing soil erosion in places. The unfortunate side-effects of hedgerow destruction have not been felt in Ireland to anything like the same degree.

10 *Male Marsh Harrier*

Herbicides, insecticides and other chemicals which have had toxic effects on wildlife in the past, particularly in the 1960s and '70s, continue to have an adverse effect, though due to helpful legislation, less significantly than formerly. The build-up of organochlorines in birds of prey, at the top of their food chain, was seriously affecting the breeding success of certain species. The crash of the Peregrine was the most publicised example. The reduction in use, and banning, of Aldrin, Dieldrin, 245–T and DDT eventually allowed the birds of prey to recover: the Peregrine to more than 200 pairs. This proved to be a valuable example of birds acting as monitors for the state of health of the countryside. Of course, sympathy for the fate of one of our diminishing stock of raptors had little to do with the action. It was self-interest – the threat of danger to ourselves, at the top rung on the ladder of life – which prompted it.

While we are now, undoubtedly on the threshold of more enlightened times, the problem continues. During the summer of 1999 two Scottish Golden Eagles were poisoned by Carbofuran, a farming insecticide, left out deliberately as a bait to kill predators.

The accelerating development of rural Ireland, particularly in formerly less-populated counties, is increasing the pressure on wildlife habitats, some of which hold threatened or endangered breeding birds. The construction of new bungalows and holiday homes along the dry hillsides and headlands of the eastern counties has encroached upon undisturbed haunts of birds like the Nightjar. Tracts of blanket bog and mountain wilderness are being steadily depleted by forestry plantations which threaten the habitats of birds like the Red Grouse and the Merlin, though others like Hen Harriers and Whinchats and Grasshopper Warblers have learned to adapt (temporarily) to the young plantations. Forestry plantations have provided alternative habitat types but they only support a fraction of the rich bird population to be found in native hardwoods and it is likely that they adversely affect the species richness of adjacent wetlands. Commonage subdivision and reclamation and the conversion of machaire and sand dune systems, formerly rough grazing, to golf courses are among the most serious contemporary threats to bird (and other wildlife) habitats.

HUNTING AND PERSECUTION

The hunting of birds is as old as mankind. Mesolithic nomads were dab hands at it, trapping a wide range of bird species for food. Trapping and hunting wild birds particularly for food continues today throughout the world. It is less evident here and in Britain than in the Mediterranean countries, where millions of mainly migrant birds are killed each year. The outrage expressed by bird groups reflects both an enlightened 'post-hunting' outlook and an indignation born of the knowledge that they are killing 'our' birds besides their own. Such concerns were not at issue in Gaelic times. Birds were trapped for food and sport, for caging and keeping as pets. The Brehon laws contained clauses concerning the distribution of the flesh and feathers of netted birds and set out the circumstances in which trapping them was legal. Though hawks were exported as rents or gifts, there is no evidence that falconry was practised prior to Norman times. The Normans were avid hunters with hawks and hounds. The sheriff of Dundalk is recorded as hunting Partridges with hawks (probably Irish Goshawks) in the early 14th century, and doubtless they were also flown against Cranes.

A series of acts of parliament since colonial times provided protection to some birds – notably game birds – but, due to the private interests of the prime movers, paid scant regard to the protection to our birdlife in general. 'Vermin' lists were not confined to crows and birds of prey. They included Cormorants, Grey Herons and other species which were perceived to be undesirable.

The numerous written accounts of the 19th-century hunters and trappers, wildfowlers and gamekeepers point to a savage disregard for birds. They were shot, trapped or otherwise killed for a variety of reasons but particularly for sport and for food. Birds could be shot at any time of the year, and little account was taken of the importance of providing undisturbed breeding conditions. The brief of the gamekeepers, who were employed by every demesne and estate throughout the country, was to exterminate any predatory species that threatened the well-being of the introduced game stocks. Naturally birds of prey were prime targets, and most other large birds were regarded as potentially undesirable for one reason or another.

It is highly likely that shooting pressure was the 'final straw' for the wetland birds which had already been seriously affected by the drainage of their habitats. Birds like the Crane and the Bittern were desirable for the table, while the Marsh Harrier was regarded as a threat to the introduced game. No group of birds was more persecuted than the birds of prey. Entries in the estate game books show that the efforts of the Victorian gamekeepers, more than others, so depleted the numbers of raptors in Ireland that it is a wonder that even the common and widespread Kestrel survived. In the early 19th century there were eleven species of birds of prey resident in Ireland (not counting the Goshawk, the Kite and the Osprey which had already gone): by the early 20th century there were only five or perhaps six. Two species, the Peregrine and the Hen Harrier, had managed to remain only by holding on in the remotest corners of the country.

Throughout the 19th century all birds with hooked beaks were classed as vermin (along with a variety of other birds and mammals) and were relentlessly shot, trapped or poisoned.

Trapping of birds was particularly effective using the pole trap. This cruel device (now banned) was simply an iron gin trap mounted on the top of a pole, a likely perch for a raptor. The bird, on landing, was gripped by the leg(s) in the trap's teeth in the manner of a rat trap, causing it to expire, not immediately as with the rat, but slowly and painfully.

Other such devices were the noose trap and the net trap which caught birds of prey with live decoys such as pigeons.

The eagles were the enemies not only of the gamekeepers but also of farmers, especially sheep farmers. The war against them was carried out single-mindedly and ruthlessly. Not only were they shot and trapped where this was possible, but eyries were robbed by paid desperadoes who risked life and limb by scaling down sheer cliff faces suspended on a rope, These forms of persecution met with varying degrees of success. In many instances the wily eagles simply built elsewhere. The widespread introduction of strychnine in the latter part of the 19th century finally sealed the fate of Ireland's eagles. Dead sheep and other livestock were doped with the poison, and birds that took carrion were killed in large numbers. Many species were reduced as a result of this activity. Foxes and other carnivorous mammals as well as Ravens and other crows were most affected. By the turn of the last century, the campaign was so successful that the Buzzard first, followed by the White-tailed and then the Golden Eagle, were completely exterminated.

To ignore the effect of egg-collecting and specimen hunting would be to omit a significant and often final chapter in the story of the exterminations. Private collections of both eggs and stuffed skins – particularly those of birds of prey – were fashionable throughout the 19th century. Rivalry existed among collectors, and no effort was spared by individuals to outdo the display of others. The success of these collectors was greatly assisted by their affluence and their capacity to pay the highest prices to those prepared to take the considerable risks.

To keep the effects in perspective, however, it must be said that collecting would not have been so important a factor in the extermination of these birds had not they become rare through other pressures. It was the fact that eggs attained high acquisition status as trophies that hastened the demise of these birds. The necessity for the lord of the manor to display a uniquely rare conservation item in a glass case was, in the final analysis, the critical factor. In response to the demand for finely displayed specimens, the art of taxidermy attained its acme in the later 19th century. There were taxidermists in the cities and larger towns, and it was customary for interesting specimens shot at home or abroad to pass through their hands. It is difficult to assess the overall affect of the work of taxidermists on the decline of some species, but it can be fairly well assumed that the

11 *Golden Eagle, immature, shot in Connemara, 1898*

sporting customers of the Sheals family in Belfast or Williams & Son in Dublin were responsible for dispatching the last of our extinct birds of prey.

Hunting has, since time immemorial, taken a toll of bird populations, but it is not generally considered to be the major cause of extinction. There are, of course, exceptions, like the Passenger Pigeon of North America or the Moa of New Zealand which were definitely hunted to oblivion, the first by waves of colonists, the second by aboriginal natives. The massive loss in the birdlife of the Pacific islands – hundreds of species of crakes and rails – is attributable to a combination of hunting, predation by introduced animals, and habitat destruction. Since these birds had evolved in an environment free of large predators and were in many cases flightless, they were 'ecologically naive' and defenceless against the onslaught of humans. In the past, the effects of hunting on birds, whose capacity to survive had been otherwise significantly diminished, have been shown to be critical.

In the case of Ireland's lost birds we can point to no single overriding cause of their disappearance. Of the influences we can isolate, however, geographic location and the biogeographical constraints on an island bird population have obviously been significant. Climate seems to have played a much lesser role. If we take the unreasonable stance of isolating the detrimental effects – direct and indirect – of human involvement, habitat destruction has been the most important and enduring factor. The majority of the lost birds discussed in the case studies felt the heavy hand of habitat destruction, especially woodland destruction. All of the lost birds of prey were adversely affected by hunting. Some, vulnerable through loss of habitat or desirable as trophies, were subsequently hunted into oblivion.

About a fifth of the world's bird species have become extinct over the past four thousand years: a fifth of those remaining have gone in the past four hundred years. This accelerated decline is symptomatic of present-day loss in biodiversity which some ecologists are already calling the 'sixth great extinction'. It may be premature to think as yet in those terms, but there is no reassurance to be found in the birds which comprise the case studies of this book, most if not all of which have been lost to Ireland in the past four centuries.

PART TWO

THE LOST BIRDS

The case studies that form the core of this book concern eleven birds which the available evidence shows to have been formerly indigenous here. There are, however, many references to other birds about which similar evidence is lacking or is so paltry that it is impossible to develop case studies. Most such references are in the form of passing comments, now difficult or impossible to substantiate.

Often these allusions raise niggling questions. Was the writer in fact referrring to the species which we now know by the name used? Was the bird mentioned widespread or simply a one-off? Could it have been introduced from another country like the Barbary ape found at ancient Emain Macha (Navanfort, Co. Armagh). How much was the comment influenced by hearsay or myth (Giraldus et al.)?

Besides that given in the case studies, there exists archaeozoological evidence of a number of species not now found here. When this cannot be corroborated by literary material (and even in one or two cases where it can), it is difficult to develop a picture of the former status of the bird in question. Even when it can be established that a particular bird was once common in Ireland and bred here as well, this may not represent the 'normal' situation; in other words the bird may not have been a long-term indigenous resident: bird populations, like those of other organisms, are constantly in a state of flux.

An example is the Grey-lag Goose. Its present status is as a winter visitor in relatively small numbers to certain favoured localities. A hundred years ago, however, thousands of Grey-lags wintered in Ireland, particularly in the flat, reclaimed land near Wexford. The Bean Goose was described (by William Thompson and contemporaries) as being widespread in Ireland and may have been the winter goose in the country at that time. This has been disputed by modern ornithologists, since the Bean Goose is only a rare straggler nowadays. However, the picture of a Bean Goose drawn by Templeton nearly two hundred years ago certainly looks like one. In the present century the Greenland white-fronted Goose has become the winter goose of Ireland, having apparently ousted the Bean, and it remains our only widespread 'grey' goose.

All this suggests a constant flux in the fortunes of our wild goose population and makes it difficult to think in terms of 'static' or 'indigenous' populations.

However, it does seem from the available evidence that one of these geese – the Grey-lag – was formerly indigenous here. Rutty (1772) stated that 'It breeds here, particularly in the Bog of Allen'. Harris (1744) also knew of the 'Great Harrow Goose' (considered to be the Grey-lag) from Kirkiston Bog (now cut away, but shown clearly on contemporary maps) on the Ards peninsula, and there are other references to its breeding there. (There is evidence that the Grey-lag bred in the English fens up to about this period.)

When we consider that the raised bogs (Kirkiston, Allen) were in the early 18th century still mainly impassable 'wastes', turf exploitation being still merely peripheral, it is easy to see how big, 'huntable' birds could find the conditions suitable to continue to breed there until that time. Though goose bones are notoriously difficult to differentiate, the quantity of fossil material from sites of early human habitations (crannogs, mainly) point to this goose being a favourite food item throughout the prehistoric and most

*Anser segetum   Bean Goose length from the bill
to the tail 27 bill to the toes 20½ inches Weight 7 pound
(shot in the Bog meadows 6 Feby
1801 by Eccles Rice)*

12 *Bean Goose by John Templeton,* c. 1800

of the historic period. Interestingly there is also bone evidence of an unidentified goose from Shandon cave, Co. Waterford, which has been dated to the glacial period (*c*.31,000 years ago). It was found amid faunal remains which included Arctic Fox, Brown Bear, and Wolf (Woodman, McCarthy and Monaghan, 1997) and could have greased the chin of any of these predators.

It seems likely that the Grey-lag Goose was an all-year-round resident in suitable wetlands throughout the country and that the allusions to it breeding in the 18th century refer to remnants of a once widespread post-glacial population. Ultimately, as in the case of the Crane which cohabited the bogs with the Grey-lag, it was simply too big, too obvious and too edible to survive the demands of a developing society.

The Black Grouse or Blackcock, a game bird well known in the north of Britain, is usually thought of as not being native to Ireland. There is no reliable report of it in historical times, and none of the many attempts at introduction has succeeded. This can be taken as 'supplementary indicative' rather than conclusive evidence: the attempted re-introduction of Capercaillie (now known to have been native once) in Sligo also failed.

The case of the Black Grouse is not entirely clear. Bird bones from the Mesolithic sites at Mount Sandel (near Coleraine) and Lough Boora included those of 'indeterminate' grouse. The archaeological report (Van Wijngaarden-Bakker, 1986) pointed to the difficulty in differentiating between the bones of Red and Black Grouse and added that 'There is some slight evidence, however, that seems to point to the Black Grouse.' Bones, once thought to be from Black Grouse, were found in a cave at Ballynamintra, Co. Waterford. Like the previously identified 'turkey bones' also found in Irish caves, these could do with being re-examined in the light of advances in our archaeozoological expertise. Scharff (1915) listed two Irish names, *Liath chearc* and *Cubaire*, and found an equivalent for the former of these names in Scots Gaelic. (The Black Grouse is still found in Scotland.) O'Sullivan Beare (1625) mentions the *Cearc dubh* and the *Coileachgdhuibh,* but he does not describe them as being specifically native here and does not give a Latin name which would help identify the species he had in mind.

Thompson (1850) gave his verdict in his *Natural history of Ireland* in which he stated: 'I have not met with any satisfactory evidence of this very fine bird having been indigenous.' Perhaps if this game bird was resident in the ancient pine or birch/heath habitat which once supported the Capercaillie, its remains may eventually come to light in the middens adjacent to early eating places for, like its larger relative, the highly edible Black Grouse would have been hard to resist.

Scharff thought that the names *Tarmochan* and *Tarmonach* were Irish names for the Ptarmigan, stating that they corresponded to words in Scots Gaelic. He also mentioned that a bone found in Shandon cave could be 'doubtfully referred' to this species. However, unless the bone dates from the immediate post-glacial period, it is difficult to imagine Ptarmigan in this country, for the conditions which support it in the Scottish highlands (mountain

13 *Capercaillie at Blackcock lek*

tundra) at the southern limit of its Arctic range no longer exist here.

Another bird mentioned by O'Sullivan Beare in his *Zoilomastix* is the White Stork, which he describes and gives both Irish and Latin names – *Corri greni, Ciconia*. It is interesting that this is the name given by others including Bedel in his late 17th-century translation of the Bible (for the Grey Heron?). H. O'Sullivan (c.1830) also gives this as a name for the heron rather than the stork. As the normal name for the heron is *Corr riasc* (or similar), could this be another example of name transference? Scharff lists *Corr bhán* and *Corra bhán* for White Stork and states that the existence of the name 'indicates that at one time they were more common here'. The name (apparently) exists in the Book of Leinster and is identical with the Scottish name. The idea of the former existence of White Storks in Ireland may not be as far-fetched as it at first seems: they are known to have nested in Britain in the Middle Ages and a pair took up residency on the steeple of St Giles' church in Edinburgh in 1416 (Gooders, 1983).

Thompson thought that the Great Bustard might have been a former Irish bird and quoted Smith's *Natural history of Cork* (1749) where this is stated to be so. Over a century earlier O'Sullivan

14 *Heads of Greenfinch, Hawfinch and Bullfinch*

Beare also mentions this bird which he specified by its Latin name *Otis tarda*: 'The bustard is bigger than the domestic cock and so heavy, weighty and slow in flight that if a hunter earnestly pursues it, in the open, he will overtake it. Dark in colour it is welcome as a pleasant food.' In the margin he writes, *'Tarda fertur esse in Ibernia'* (The Bustard is said to be in Ireland). Though unlikely in Ireland today, this may not have always been so, since, in the absence of any protection, an enormous game bird such as a bustard would be an obvious target for hunters. At one time (up to the beginning of the Early Modern period) the Great Bustard was, by all accounts, a locally abundant resident in a scattering of locations in Britain. It was known from open country in Berkshire, Herefordshire, Cambridgeshire, the Wolds of Lincolnshire, the Downs of Sussex, as well as several places in the west country. It was found in 'droves' in the Yorkshire Wolds and was finally eradicated from Norfolk and Suffolk (by egg-collectors) in 1838. Nor was it restricted to England. Hector Boethius in his *History of Scotia* (1526) stated that it bred in the lowlands of Merse in Berwick, and it was last shot in Scotland as late as 1830. O'Sullivan Beare does not proffer an Irish name for the bird; it would seem that, if indeed it was found here at all, it was not widespread.

Smith also mentions the Stone Curlew as being native to Co. Cork in his time and provides a convincing description: 'The Stone curlew runs very swift on the sands and stops all at once (say the fowlers) making the least motion with their eyes, much less of their bodies … Its feathers and feet resemble those of a bustard, and its cry is something like that of a Green plover [Lapwing]. We have it on our shores … ' This bird was unknown to other writers (apart from as a rare straggler) and its former status must remain a mystery on the basis of the evidence. However, it is still found as a rare resident on the south coast of England, having declined from a much larger range in the past. It is one of a suite of Continental birds which is seen as having declined as a consequence of climatic change (and subsidiary factors). There is a possibility therefore, that in earlier, milder times birds such as Stone Curlews, bustards and even storks might also have been found in the south of Ireland. Pelican bones dating from the Roman period have been uncovered from Glastonbury in England (cf. the notes of W.H. Hudson, 1894). Could it be that at the equivalent period – the Early Medieval – we were also graced by this magnificent wetland bird? Gwithers, who submitted natural history notes to the Dublin Society (*c*.1684) included the Pelican and the bustard under 'Birds found in England but not in Ireland'. Under 'Birds found in Ireland, not found in England' he listed the [Golden] Eagle and the Capercaillie (Cock of ye wood: Urogall) – both probably accurate – but what was the 'Irish nightingale' and, more enigmatically, could there really have been a 'Great Irish Owl' not now found here? If (as seems unlikely) Gwithers was accurate on the latter score, he could really only have been referring to the Eagle Owl, that massive bird of prey of the European forests. There is no corroborative evidence from other literary sources. However, its former existence here, in more afforested times, is certainly within the bounds of possibility. Who knows what bone evidence might eventually reveal?

The smaller birds are much more difficult to account for historically. They were much less noticed than large birds and were often ignored unless they had some food or other value and were readily caught. In addition, the names of smaller birds have been interchanged more often than those of larger birds, making identity more difficult. Etymological work in English has helped to clarify archaic names such as 'pettychaps' for warblers and so on. In Irish, however, the problem is greater in that quite unrelated small birds have been grouped together, often because of one striking feature or other. A forked tail has spawned the name *Gabhlán gaoithe* meaning 'forked tail of the wind'. The name has been applied to birds as unrelated as the Swallow and the kite which have virtually nothing else in common.

In the case of some small birds the name is unambiguous. The name *Deargán alt*, listed by Scharff (for which he stated an equivalent in Scots Gaelic) can really only be the Redstart. Nowadays the Redstart is very local and confined mainly to oakwoods. The occurrence of the name in several lists would suggest that the bird was more well known in the past, presumably in the days before the final woodland clearances.

The undated fossil remains of the Hawfinch (including the massive lower mandible) were found in cave deposits in Co. Clare. They were found alongside those of other silvan species – Great-spotted Woodpecker, Jay and Sparrowhawk – and suggest an avifauna richer than that found in Clare today. Nowadays the Hawfinch is a rare visitor to Ireland, usually solitary, but in an unprecedented influx in 1988–9 over a hundred spent the winter feeding on the ample supply of nuts and fruits in Curragh Chase, Co. Limerick. It is interesting also that two adult Hawfinches were seen with a young bird, obviously bred locally, at Ballyvaughan, Co. Clare, in late summer 1991 (P. & J. Clark). This one-off nesting was probably due a pair from the winter influx lingering on after the flock had returned (to Europe?) but it shows a propensity for Hawfinches to nest here at least on a sporadic basis. In this way it reflects the more regular incursions of the Crossbill. It is a fair bet that more evidence of the former occupation of Ireland of woodland passerines such as Hawfinches and Crossbills will come to light from future archaeological investigation.

The Woodlark has been lost to Ireland within recent history. Smith and Rutty knew of it in Munster and Leinster in the mid 18th century, and Templeton, who wrote most about it, knew of it as a common bird in the environs of Belfast some two hundred years ago. It was thought to have nested in favoured localities in Ulster,

15 *Woodlark at nest*

Leinster and Munster until the mid 19th century: its range included counties Antrim, Down, Armagh, Dublin, Wicklow, Wexford, Waterford, Cork and perhaps one or two others. It declined noticeably towards the end of the 19th century by which time it remained only in Down and Wicklow. It had vanished by the start of the new century. Nowadays it is only a very rare visitor here.

Ussher stated that its disappearance was due to the activities of bird-catchers; the Woodlark was much sought after as a caged songster. Liming of twigs with a gummy substance (usually holly sap) was the common practice. Thompson also alluded to the birdcatchers who were ready and growing suppliers of songsters for the sentimental tastes of the occupants of the 'big houses'. Templeton, who noted on one occasion six woodlarks singing near his house, was also aware of the activities of the bird-catchers. He wrote: 'This bird [the Woodlark], although next in mellowness and plaintive notes to the Nightingale, is but little known and wantonly destroyed by country boys during snow; it is much more local than the Skylark, being confined to districts abounding in gently rising hills, which it enlivens with its song for nearly eight months, and even by moonlight.'

16 *Spotted Crake by Dunscombe Parker, c.1840*

The disappearance of the Woodlark may have been triggered by climatic change since it has been shown to be subject to massive population fluctuations as a consequence of cold winters. It has been suggested that in conditions of mild winters the Woodlark – a central European species at the edge of its range in Ireland – does not take the partial migration option and remains throughout the year near the nest site. A sudden cold winter (or series thereof) can thus decimate a peripheral population. Despite the observations of Templeton et al. the bird-catchers may have been 'mopping up' an already, severely depleted stock.

There has been some speculation about the possible breeding of the Tree Pipit in Ireland. Hutchinson (1989) thought that, due to

the number of sightings recorded from south coast observatories, a few probably nested unnoticed each year in Ireland. Sharrock (1969) queried whether it formerly nested in Ireland and had perhaps died out. Its absence from Ireland on the breeding atlas is striking, given that it now breeds throughout Britain from the south of England to the north of Scotland. It is however a relative newcomer to Scotland, having spread up initially to the hillside birch woods. The extensification of conifer plantations since the war has consolidated its spread. The equivalent recent plantation programme in Ireland has not had the same effect. It seems likely therefore that the Tree Pipit was not a former widespread inhabitant of this country, that factors such as dampness or simply our geographical location have militated against it and that those which turn up regularly on migration are wanderers from Welsh or Scottish breeding sites.

One small bird which once nested here and has disappeared in the present century is the Yellow Wagtail. It may never have been widespread here as it is over southern Britain and much of Europe but it was found in certain habitat – wet pasture and callow – in a number of widely separate localities until the middle of the 20th century. In the 19th century it nested colonially on callows at the north and south ends of Lough Neagh and on lake islands and shores of Loughs Corrib, Mask and Carra in Connaught. There was (and indeed still is) sporadic colonial nesting in suitable habitats along the east coast.

 No one can say now if the Yellow Wagtail was, like the Tree Sparrow, a relatively recent 'invader' which for reasons unknown arrived in the country a century or two ago and again for reasons unknown subsequently steadily declined and eventually died out completely. Or was it a relict – a species that perhaps, due to the mid 19th-century drainage of wetlands by the Commissioners of Works, or changing climatic conditions, simply declined and disappeared. Neither earlier literary nor archaeozoological evidence can clarify this for us now. However, in view of the fact that large tracts of callow remain much as they have for centuries – around Loughs Neagh, Corrib, along the Shannon etc., it is hard to believe that wetland destruction – extensive as it has been – was the main cause.

The same does not apply in the case of the beautiful Black-necked Grebe for it almost certainly died out as a consequence of hydrological changes. Prior to the amazing discovery of a substantial breeding colony (perhaps 300 pairs!) in the early 1930s in a turlough/lake in Co. Roscommon, this grebe was known only from one or two previous breeding records and as a rare winter visitor. It was subsequently found nesting elsewhere in similar ephemeral lakes, but disturbance, predation by rats, Hooded Crows etc. and hydrological interference are thought to have rendered these sites unsuitable. The main site at Lough Funshinagh became unsuitable for the grebes from the mid 1930s

on: Humphreys (*Irish Birds*, 1978) surmised that a series of dry summers had a bearing on the decline of the colony but hinted that human interference – opening swallow holes to facilitate drainage, and the construction of the Shannon hydroelectric scheme in 1934 – could also have been involved. Small numbers of the grebes continued to nest at the lake, when conditions allowed, for years afterwards, but eventually they died out. It has had a similarly disastrous history in Britain.

There can be no doubt that the Spotted Crake, wetland counterpart of the Corncrake and close relative of the Water Rail, was much commoner in the past. Prior to the drainage works of the mid 19th century it was probably a widespread breeder in the sedgy wildernesses of many Irish counties. While we know nothing of it from bone evidence, there are a number of references which point to it being much better known a century or two ago than at the present time. Sampson (1802) in the *Statistical survey of Co. Londonderry* mentions the 'Spotted water-hen' and confirms that he means the Spotted crake by adding the Latin name *Porzana*. McSkimmin (1811) also refers to the 'Spotted water-hen'. Templeton and Thompson were familiar with the Spotted Crake from hunters' 'bags' in the north of Ireland, while Harvey and Watters referred to it from the south. Harvey in his *Fauna of Cork* (1845) described it as 'more common than is generally supposed', and Watters, who called it the 'Little Crake' (another European bird entirely), knew of it from Co. Roscommon in particular, where it was proved breeding in 1856.

There are many records of bird 'obtained' throughout the latter half of the 19th century, the majority in autumn though one or two in the breeding season (Co. Laois, 1880; Fermanagh, 1890). Many of those shot in and around the north came under the notice of the famous Belfast taxidermist family, the Sheals, who stuffed them for their clients. A report in the *Irish Naturalists Journal* in 1892 from a H.D.M. Barton of Co. Antrim is most revealing: 'The Spotted crake (*Porzana maruetta*) [since changed to *Porzana porzana*] in County Louth … The district lying inland from Dundalk on the one side to Crossmaglen and on the other to Ardee abounds in marshes … Many of these, even to the most experienced snipe-shot, are almost inaccessible, being composed of a floating sedge which a dog can hardly cross ... In the same district I secured specimens of the Spotted crake. The latter [Spotted Crake] was found frequently, but although I have no doubt that they breed in the district, I was never successful in my search for their nests. Like the Water rail and indeed all the gallinules, the Spotted crake rarely resorts to flight unless there is no other means of escape, preferring rather to take refuge in the nearest tuft of grass or sedge until found by the sportsman's dog. When it rises its flight is rapid, not unlike that of the Quail and being but a small mark is not easily shot ...' It should be remembered that this was written at a time when there was virtually no restriction on hunting and when the normal way of identifying rare birds was to shoot them. It is worth noting also that these were pre-arterial drainage days and that nowadays only a trace of these once extensive wetlands remain.

Unfortunately we do not have a descriptive overview of the birdlife inhabiting Ireland's pre-drainage wetlands as we do for those of Lincolnshire from the pen of Pennant in 1770. These fens have been likened to the 18th-century fens of the Irish midlands which, being of a similar extent, must have harboured a similarly rich aquatic avifauna. Pennant wrote:

'The birds which inhabit the fens are very numerous: I never met a finer field for the zoologist to range in. Besides the wild duck ... wild geese, garganies, pochards, shovelers and teals breed here ... pewit gulls [Black-headed Gulls?] and Black terns abound ... the black and dusky grebe [Black-necked Grebe?] and little grebe are also inhabitants of the fens together with coots, waterhens, spotted water-hens [Spotted Crakes] water rails, ruffs, redshanks, lapwings ...'

Most of the amazing birdlife of the Lincolnshire fens did not survive the 19th-century drainage, and it is hard to believe that a century of drainage did not have the same effect in Ireland.*

---

* In 1994 at least seven Spotted Crakes were identified (by their singular 'whip-lash' call at night) during the summer at one or two midlands localities. Several were heard in the marshy floodplain of the Little Brosna river (Offaly/Tipperary border). Could it be that this elusive aquatic crake continues to breed here despite the widespread loss of its primary habitat?

BITTERN

Botaurus stellaris

The Bittern is a member of the heron family of which only one, the Grey Heron, is common in Ireland. It is smaller but is still a large bird, 2.5 feet (0.76 m) long. While the Heron appears generally grey, the Bittern looks generally golden brown. The plumage is heavily barred, streaked and otherwise marked with dark brown and yellowish – hence the Irish, *An Bonnán Buí,* the yellow Bittern. In flight, the profile is like that of the larger bird with the tucked-back head, protruding feet and ponderous wing beats.

For so substantial a bird the Bittern is extremely elusive and is rarely seen in the normal course of events. It usually remains motionless in the midst of aquatic vegetation relying on its cryptic plumage as camouflage. On being disturbed, it typically stretches its neck skywards to merge marvellously into its reedy background. It is decidedly reluctant to fly. In the breeding season where more than one pair are occupying a reedbed the males may indulge in aerial skirmishes, in the fashion of Grey Herons. Most sightings are, however, towards sundown when it may be seen flying to and from a favoured feeding area or going to roost.

The Bittern inhabits freshwater wetlands throughout much of temperate Europe. The primary habitat is extensive reedbeds, though it may be found in other wet locations outside the breeding season. In western Europe it is mainly a resident species though migration from the east occurs in winter. Indeed, the Bittern is very susceptible to cold winters (see the poem quoted in the end Note), and many which forgo the option to migrate die off as a result. Since it disappeared as a breeding bird in Ireland, the vast majority of its random occurrences have been in the winter.

Bitterns are specific in regard to their breeding requirements. They need extensive fen vegetation – reedbed, sedge etc. – which is relatively inaccessible and undisturbed. The nest is invariably well hidden, usually in the heart of a reed swamp or similar inaccessible morass with high emergent vegetation. Five eggs are normally laid in April or May, and the young often leave the nest within a few days of hatching.

Bitterns also require open water, drainage ditches and pools, in which they can feed. The main food is coarse fish, especially eels (on average almost 50% of their food is eels), though they eat a large variety of other aquatic animals (including rats) and insects besides.

A secretive and solitary bird, the Bittern would go largely unnoticed were it not for its far-carrying booming call, in fact, the song of the male. The sound, a 'dismal hollow boom', is low and eerie, like a distant foghorn. Emanating as it usually does from an inaccessible mire and most regularly at dawn and dusk, it has spawned a wealth of folklore and fable in Ireland, as elsewhere.

The supernatural quality of the noise captured (according to Goldsmith) the imagination of both poet and peasant, inspiring in the former fascination; in the latter, fear. Chaucer, Dryden, Thompson, Elliott, White, Shelley, Scott, Wordsworth and Ledwidge are among those who alluded to the sound in a poetic context. Ledwidge, however, is unlikely to have actually heard it, since the Bittern was long gone as a breeding bird from Ireland by his time (in the early 20th century). Mac Ghiolla Ghunna's celebrated early 18th-century poem 'An Bonnán Buí' (The Yellow Bittern) while not alluding to the bird's boom (the bird is dead) certainly assumes a widespread familiarity with the Bittern's appearance. Goldsmith, in his *Animated Nature,* wrote of the trepidation experienced by the poor Irish, who regarded the noise as a portent of some sad event, perhaps the demise of the listener. The weirdness of the sound was used to create dramatic effect in literature. In Conan Doyle's *Hound of the Baskervilles,* Watson assured the startled but intrepid Sherlock Holmes that an unearthly sound they had both heard across the moor was made by a Bittern. There is a similar passing reference in *Hereward the Wake.*

The variety of names which it has acquired down the ages throughout Europe testifies to the interest evoked by this bird. It was known in Greece as *tarrabusso* (bullroarer) or *terrabus* (earthroarer) and in Italy as *trombono,* with obvious connotations. In 15th-century Germany it was thought to make its boom by immersing its beak in water. In 16th-century England folklore had it that the sound was magnified by a hollow reed (Turner, 1544). The old names in English were also descriptive. The oldest known is *myre drommel* or *mirdrommel,* which was corrupted to miredrum, the name by which it was known to Goldsmith in 18th-century Ireland. Night raven was another ancient, colloquial name. Bittern originates from the French *Butor,* known from the 12th century: it derives from a much earlier Latin source word *But* (the sound) combined with *tor,* the Old French word for bull (Lockwood, 1984).

In the Irish language Bittern is *bunnan* or *bonnan,* often with an appended adjective, as in the poem, 'An Bonnán Buí'. Other Irish names were probably local variations and are of minor importance. Interestingly there is also a name, *Tarbh curraig* (Ó hAonghusa, 1940) which is redolent of the European 'bull' connotation.

There are several placenames called after the Bittern. These are found in a number of counties and are usually associated with some marsh or bog as in Curraghbonaun in Sligo (the marsh of the Bittern) or Feabunaun in west Cork (the marshy stream of the Bittern). In some cases the marsh or river has been long drained and it is difficult to imagine the former connection (Joyce, 1900). In other locations (Tievebunnan near Boho, and Inishbobunnan) the topography is untypical and the association obscure.

The commonest appendage is undoubtedly *léana,* meaning 'swampy expanse', and this is the name given by O'Sullivan Beare in the early 17th century, by K'Eogh in the early 18th century, by O'Suilleabhain in the early 19th and in most of the 20th century lists. Nor has it completely disappeared. *Bonnán léana* was the name given to a 'drumming' snipe by an old man near Tuam, Co. Galway, in 1985. The name, still applied to a strange bird-sound emanating from a marsh, has, in the absence of its original maker, been mistakenly transferred to the Snipe. There are many instances of name transference such as this in the Irish language (*per* de Bhaldraithe, RIA).

Bittern bones have not yet been specifically identified from Irish archaeological sites. In view of increasing emphasis on wetland sites, however, it is likely that they will turn up among the midden refuse with other bird remains.

In the many sites in Britain and elsewhere its bones have been dated back as far as the Neolithic. They have been identified in fen peat deposits on the Isle of Ely in England, where it is considered to have had a long and continuous occupancy.

The Bittern has, for centuries, been a favoured bird for the pot. It was on the menu in innumerable 14th-century dietaries of English estate houses. On a price list of this period the Bittern was itemised at 18*d.,* 2*d.* higher than the bigger and meatier Grey Heron. Mention is made of the Bittern being 'not so hard of digestion as the Heron' and it was recommended that 'it should be eaten with no sauce, only salt' (Dietary notes of Hunstanton estate, 1542).

In Ireland too the Bittern was a delicacy but only among the gentry. It was written about as follows by Goldsmith: 'Whatever terror the boom of the Bittern may inspire among the simple [folk], its flesh is greatly in esteem among the luxurious. For this reason it is as eagerly sought after by the fowler as it is shunned by the peasant.' Furthermore, 'The Bittern has become more scarce than formerly since its flesh has been accounted a great delicacy. Poulterers value it at half a guinea.'

Even as late as the early 19th century it was recounted by Maxwell in his *Wild sports of the West* that an otter hunter received

17 '*Bittern and Partridge*' *by Jean-Baptiste Oudry, 1747*

many a half crown for going to Bishop Beresford with game and fish – on one occasion with a haunch of red deer and a Bittern. A number of wildfowling pictures depict the Bittern as a quarry species. The well-known Malton prints illustrating Irish Hunting Scenes of the 18th and 19th centuries show, on one titled 'The Water', a wildfowler shooting at waterfowl with an easily recognisable Bittern hanging from his waist string. The 18th-century still-life painting *Butor et Perdix* (Bittern and Partridge) by the French artist Oudry shows these birds hung up for the pot. The Bittern was also featured in a painting by the now-forgotten Irish painter, Charles Collins; it was shown in Dublin in 1735.

Bitterns were sought after for other reasons too. In *Zoologica Medicalis Hiberniae* (1739) they were stated by K'Eogh as having therapeutical qualities: 'The flesh pulverized cures a Dysentery, Colic and Griping of the guts being taken for a considerable time, in some proper vehicle. The fat or oil is useful to cure pains, aches, contractions of the sinews and the cold gout, the grieved parts being rubbed therewith. The ashes of the feathers stop bleeding.' The feathers had another use besides. A certain type of artificial fly, used for catching salmon, was made from the feathers of a Bittern (clearly marked 'Bittern hackles') but is also simply called a 'Bittern'. It would be no easy task to obtain Bittern feathers for this function today. It is unlikely that Bitterns were actually hunted for their feathers: they were surely taken from available carcasses, birds shot for food. This applied also to the Corncrake, the carcasses of which supplied feathers for 'Rail flies' (Fox & Sleeman, 1995). However, as the authors point out it is unlikely that the practice actually threatened the Corncrake population and the same can be said of the Bittern.

As Bitterns became rarer, in the 19th century, they became sought-after trophies for stuffing and displaying in a prominent place in the 'Big House'. Thomas Sheals, the famous 19th-century Belfast taxidermist, noted that, prior to keeping records of specimens received, almost every year a dead Bittern or two was brought for mounting to his shop in Corporation Street. In an interesting anecdote he draws the reader's attention to the dangers of keeping wounded Bitterns, stating that he had actually been attacked by one.

That the Bittern was formerly a common species in Britain and Ireland is without doubt. The references to its abundance in Britain go back as far as Bede, the monk who mentioned the existence of innumerable Bitterns in the Fens region of East Anglia in the 8th century. The dictionary of Old and Middle Irish words contain references to the Bittern of undoubted antiquity. Even as late as the 18th century it was a familiar bird to the writers on natural history, who gave no indication of it being on the decline.

By the early 19th century, however, the commentaries invariably mention a decrease in numbers and range. Man-induced pressures are consistently cited as the reason. William Thompson was a witness to the decline, and his remarks are endorsed by later writers. The following are some retrospective accounts of Bitterns collated by Ussher & Warren (1900): 'An old man in County Waterford told me in 1856 that in his youth he used to hear the sound made by these birds near the confluence of the Blackwater and the Bride, before the lowlands were embanked and that the birds bred there'; 'In Eastern Galway the late Lord Clonbrock told me of bitterns he used to meet … on Crit bog from 1819, the first year he began to shoot, until about 1830'; 'At Mantua, County Roscommon, the late Oliver Grace in the 1840s or earlier used to caution sporting visitors not to shoot the Bitterns which then bred there'; 'In Cork … [a man] informed me that his father, when

young [early 19th century] used to hear bitterns in the bog there, long since drained.'

Thompson, recognising that the Bittern was on the way out, documented its shrinking range accurately. He knew of it from nineteen counties but could only state with certainty that it bred in at least half of these. Some of his information was anecdotal. Concerning Donegal he remarked, 'I am informed that Bitterns were very common in this county thirty years ago [*c*.1810]: from increased cultivation and population they are now, however, only rarely to be seen.' He also mentioned that Bitterns were sold as food at the markets in Ennis (Co. Clare) and Dublin. A separate note in the *Magazine of Natural History* (1831) referring to the neighbourhood of

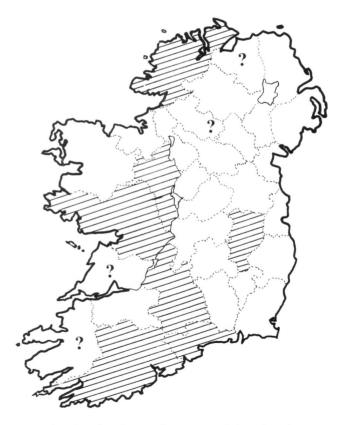

18 *Bittern breeding distribution (known counties), early 19th-century*

Kilrea, Co. Derry/Londonderry, stated that although the Bittern was resident there at that time, it was 'not common'.

From the mid 19th century onwards information regarding breeding Bitterns in Ireland was generally lacking, though it may have continued to do so in a few less disturbed haunts. It remained a widespread and regular winter visitor to wetlands throughout Ireland until the latter part of the 19th century. By 1900 the only county in which Ussher knew it to occur annually was Laois (presumably in the vast Granston fen at Durrow, still undrained then). The various natural history journals recorded Bitterns as scarce or rare winter visitors well into the 20th century.

Today it is regarded as a rare winter visitor, one or two being recorded in most winters.

The literature makes it clear that the Bittern was once a widespread and well-known breeding bird in Ireland. Taking account of the many references in English and Irish, the placenames and other more oblique allusions, we can safely say that it occurred in all four provinces and probably in the majority of counties before the early 19th-century decline. Thompson gives us our most complete historical inventory, but it is clear from the other writers that it was once a typical and familiar bird of suitable habitat.

There can be no doubt that habitat destruction and hunting both played a part in the disappearance of the Bittern from Ireland. Writing of its disappearance from Britain, Gurney stated, 'The factor which banished it from Britain was partly drainage but still more fatal than drainage has been the gun.' Thompson, on the other hand, wrote, 'Once common in Ireland, it is gradually becoming scarce owing to the drainage of the bogs and marshes.'

Already under pressure by being hunted for food (a pressure which is likely to have increased due to the rise in population and the impending famine) it does seem that the wide-scale drainage operations under the 1842 Drainage Act was the factor that hastened the decline and ensured that Bitterns would not have extensive, undisturbed wetlands to which they might return. Since these works were carried out in the summer when the water level was low and by large (presumably noisy) gangs of men, they must have had a dramatic effect on sensitive breeding species like the Bittern.

19 *Head of Bittern*

There still exist extensive marshes and fens capable of supporting breeding Bitterns but, unless conservation practices are improved in combination with a sympathetic public attitude, repopulation by natural spread from Britain seems unlikely. The Bittern remains a scarce breeding bird in Britain but due to rigid protection seems to be holding its own. Stragglers occasionally occur here in suitable habitat, occasionally even during the breeding season, (Co. Clare, *c.*1990) but we will probably have to wait for them to become better re-established in north-western Europe and particularly in Britain before we can realistically look forward to their return.

NOTE

Herewith the text of Cathal Buí Mac Ghiolla Ghunna's 'An Bonnán Buí' (The Yellow Bittern), early 18th-century followed by a free translation by Thomas McDonagh of the same poem

A bhonnáin bhuí, is é mo chrá do luí
is do chnámha críon tar éis a gcreim,
is chan díobháil bídh ach easpa dí
d'fhág tú 'do luí ar chúl do chinn;
is measa liom féin ná scrios na Traí
thú bheith sínte ar leacaibh lom,
is nach ndearna tú díth ná dolaidh is tír
is nárbh fhearr leat fíon ná uisce poill.

Is a bhonnáin álainn, mo mhíle crá
do chúl ar lár amuigh insa tslí,
is gur moch gach lá a chluininn do ghráig
ar an láib agus tú ag ól na dí;
is é an ní adeir cách le do dhearthair Cathal

go bhfaighidh mé bás mar súd, más fíor;
ní hamhlaidh atá – súd an préachán breá
chuaigh a dh'éag ar ball, gan aon bhraon dí.

A bhonnáin óig, is é mo mhíle brón
thú bheith romham i measc na dtom,
is na lucha móra ag triall chun do thórraimh
ag déanamh spóirt is pléisiúir ann;
dá gcuirfeá scéala in am fá mo dhéinse
go raibh tú i ngéibheann nó i mbroid fá dheoch,
do bhrisfinn béim ar an loch sin Vesey
a fhliuchfadh do bhéal is do chorp isteach.

Ni hé bhur n-éanlaith atá mise ag éagnach
an lon, an smaolach, ná an chorr ghlas –
ach mo bhonnin buí a bhí lán den chroí,
is gur cosúil liom féin é ina ghné is a dhath;
bhiodh sé choíche ag síoról na dí,
agus deir na daoine go mbím mar sin seal,
is níl deor dá bhfaighead nach ligfead síos
ar eagla go bhfaighinnse bás den tart.

Dúirt mo stór liom ligean den ól
nó nach mbeinnse beo ach seal beag gearr,
ach dúirt mé léi go dtug sé bréag
is gurbh fhaide mo shaolsa an deoch úd a fháil;
nach bhfaca sibh éan an phíobáin réidh
a chuaigh a dh'éag den tart ar ball? –
a chomharsain chléibh, fliuchaidh bhur mbéal,
óir chan fhaigheann sibh braon i ndiaidh bhur mbáis.

The yellow bittern that never broke out
In a drinking-bout, might well have drunk;

His bones are thrown on a naked stone
Where he lived alone like a hermit monk.
O yellow bittern! I pity your lot,
Though they say that a sot like myself is curst
I was sober a while, but I'll drink and be wise
For fear I should die in the end of thirst.

Its not for the common birds that I'd mourn,
The blackbird, the corncrake or the crane,
But for the bittern that's shy and apart
And drinks in the marsh from the tone bag-drain,
Oh! If I had known you were near your death,
While my breath held out I'd have run to you,
Till a splash from the Lake of the Son of the Bird
Your soul would have stirred and waked anew.

My darling told me to drink no more
Or my life would be o'er in a little short while;
But I told her 'tis drink gives me health and strength,
And will lengthen my road by many a mile.
You see how the bird of the long smooth neck,
Could get his death from the thirst at last
Come, son of my soul, and drain your cup,
You'll get no sup when your life is past.

In a wintering island by Constantine's halls,
A bittern calls from a wineless place,
And tells me that hither he cannot come
Till the summer is here and the sunny days.
When he crosses the stream there and wings o'er the sea,
Then a fear comes to me he may fail in his flight –
Well, the milk and the ale are drunk every drop,
And a dram won't stop our thirst this night.

RED KITE

Milvus milvus

The effortless gliding of this large bird of prey was the model for the familiar toy kite. Indeed, one of the ancient names of the kite was *glead* or *gled* which means literally 'glider', and anyone who knows this bird would testify to the aptness of the name. The Red Kite is indeed glider *extraordinaire*: it surveys the countryside from the air on long, finger-ended wings and deeply forked tail which is fanned and twisted with consummate subtlety to adjust the bird's position in the air. The profile from below is unmistakable.

The overall plumage is red-brown, particularly on the tail. The head is normally very pale with noticeable dark streaks, and there are strikingly pale patches on the underwings.

It is about 2 feet (0.6 m) long with a wingspan of about 5 feet (1.5 m).

The main prey is small mammals and birds, but a variety of other prey species is taken when the opportunity arises. Carrion is an important dietary item. Once a well known refuse collector – even in towns and cities – the Red Kite also hunted (mainly young) game and poultry in the country. While it was generally tolerated on the former score, it was despised on the latter. On much of the European mainland the Black Kite occupies the scavenging niche once held by the Red Kite, the latter having now retreated from its former widespread range. It is debatable, however, whether the Black has ousted the Red, given the former's preference for river and riparian habitats over woodland and open country.

The primary habitat is deciduous woodland, though reintroduced kites in Scotland breed in mixed woods and even in mature conifer stands. In Wales, where a relict population survives today, the habitat is oak-clad hillside and adjacent agricultural land. The breeding population is concentrated in a relatively small area, but in normal circumstances this is not the case: each pair may occupy a range comprised of woodland and open ground, extending to 4000 acres (*c.*1500 ha).

Most north European Red Kites are at least partially migratory, the majority wintering in the Mediterranean region. Those from Spain and southern Europe are mostly sedentary.

Like the Bittern, the kite caught the imagination of poets down the ages, not for its distinctive cry (though some 19th-century poets make mention of this) but for its singular appearance in flight. Its long-winged and long, forked-tail profile would not go unnoticed. Wordsworth describes this in 'An Evening Walk', as do Cowper and Bloomfield, two now obscure, late 18th-century English poets. Shakespeare mentions the kite among his many references to birds, and his contemporary, Spenser, does so too. Spenser's allusion is particularly interesting since he lived for a

20 *Head of Red Kite*

considerable time in his Kilcolman estate, in Cork, where he is reputed to have written *The Faerie Queene:* his 16th-century reference to the kite therefore could well be to the Irish Red Kite.

Bones of the Red Kite have been recovered from excavations at Wood Quay Viking site in Dublin. They date from the 10th-11th century. They were found in 'back-yard' situations along with the remains of other predatory and scavenging birds such as Ravens, other crows, Buzzards, harriers and even White-tailed Eagles. The list echoes that from Yorvic (York) and suggests that these Viking settlements were beset by a formidable collection of avian refuse collectors.

Excavations at Lough Gur in Co. Limerick have revealed bird bones from the 13th or 14th century. Here too kite bones were uncovered along with those of Buzzards, an unidentified eagle and numerous Ravens and other crows. In the archaeological report, Cleary (1982) thought that these birds might have been killed and hung up as a deterrent to other predators – much as grain farmers still do today, with crows. However, the earlier evidence suggests a more symbiotic relationship. The Red Kite bones found during archaeological work at Roscrea Castle, Co. Tipperary, date from the 17th century. The birds may have come to grief as quarry for falconers. There are many references from England and elsewhere to kites being used for this purpose.

The table of Irish names which can be associated with the kite illustrates well the lack of standardisation in regard to the names of creatures in general (see the Note below). It would be impossible now to regionalise the various names – if that were ever possible – for it is quite likely that some birds were called by a number of names in a single locality. However, as is usually the case, one name was more frequently recorded than others. *Préachán* literally means 'crow', but with an appendage like *na gcearc* it comes closer to 'hunter or predator of the hen'. This could simply be a generic name for any predatory bird which would take a hen. However, this habit has been specifically associated with the kite and points clearly to this bird over others and represents a typical Irish descriptive name (like *pocaire na gaoithe*, the wonderfully descriptive name for the Kestrel, literally 'beater of the wind').

Another name which appears to have been applied to the kite is *cromán*, either by itself or together with an appendage like those used with *Préachán*. *Cromán* literally means 'hooked' and probably refers to the hooked beak or talons of this bird. Carrickacroman in Co. Cavan is called the rock of the kite (Joyce). There are other names in the early literature such as the Brehon law-texts which may have been applied to the kite (see the Note).

In his *Topographia Hiberniae* (12th-century) Giraldus states that 'Eagles are as numerous here [Ireland] as Kites are in other countries.' This statement, if it is to be taken at face value, substantiates neither the presence nor the absence of kites in Ireland (though we know them to have been present then, from the archaeological evidence). It suggests that they may not have been as common as Giraldus was used to seeing them elsewhere – in Wales and England. They were certainly common in England, Scotland and Wales in the Middle Ages (Harting) and given the suitability and extent of the woodland habitat in Ireland at that time there is no reason to believe that they were any less common or widespread here.

Kites were common scavengers in the towns and cities of Europe throughout the Middle Ages. They were particularly common in the streets of London into the later Middle Ages, where the Venetian ambassador wrote in his diary, 'Kites are so tame that they often take out of the hands of the little children the bread smeared with butter given to them by their mothers' (Capello, 1496–7). I could not find a similar reference from medieval Dublin, but there is a brief mention from Cork city in 1621. In a remarkable though probably exaggerated account of a memorable storm in that year, a large number of starlings are said to have perished along with a Raven, a crow and a kite. Another rare allusion to the kite comes from the pen of a Captain Trevor who, in describing the famine which followed Mountjoy's Ulster campaign (*c.*1603) remarked (somewhat dispassionately) that 'horses were killed for food and not only horses but dogs, cats, hawks, kites and other birds of prey … ' (Colles, 1919).

The most complete description of the kite in Ireland at this period comes from O'Sullivan Beare, who despite intermittent moralising and fantasising (typical of writers on natural history

G. D'ARCY '85

prior to the 18th-century scientific revolution) gives us the following: '[it] does not dare to attack hens, but goes for their chicks which it ambushes suspended in the air and rarely flapping its extended wings. Between intervals of flying it guides itself, like the rudder of a ship with frequent movements of its tail. But while the thief itself circles around the weak little animals it is often assailed by the high-flying plunderer [an allusion to another raptor], against whom, however, it sometimes defends itself, fighting on its back with its talons and its beak.' Apart from the proviso on the hen, which it has been known to take when chicks are not available, this is a lucid and accurate description of kite behaviour.

There is a passing reference to the kite in *The experienced huntsman* (Stringer, 1714) where it is mentioned along with crows as scavengers of fish left behind by an otter. The location was a large estate on the Co. Antrim side of Lough Neagh. There can be no doubt that the Red Kite was indeed the bird intended by Stringer, since he was a highly competent gamekeeper.

Irish statutes of the 17th and 18th centuries of Queen Anne, George II and George III offer bounties for the killing of various predatory animals and birds. Under 'vermin' (27 Geo. III, *c*. 35) there is reference to 'Rewards for killing an otter, or marten, wasel [stoat], cormorant or kite, scalcrow [Hooded Crow] or magpie or rat.' Thompson (1849) was sceptical about the statute references and was unconvinced about the kite as a widespread Irish bird, even though two of his 18th-century natural history predecessors, Smith and Rutty, were, emphatically. In his *Antient and present state of Cork* (1750) Smith describes it as 'common' under the name '*milvous cauda forcipata*' and 'needing no particular description' but says that it is distinguished from all other birds of prey by its forked tail. Interestingly also, he states that 'It remains with us all the year.' Smith endorses this remark in *The antient and present state of Waterford* (1756) where he lists it among 'Sorts of birds common to other parts of the kingdom [Ireland], as the kite, the Buzzard'.

Rutty mentions that the Buzzard, the 'Kite or Glead' and the 'Moor Buzzard' (Marsh Harrier) were known to him in his *Natural history of the county of Dublin* (1772). He also differentiates between the Buzzard and the kite in the question of food, stating that the Buzzard hunts young rabbits and hares while the kite prefers young chickens and goslings.

K'Eogh (1739) mentioned that the kite was an Irish bird with medicinal qualities. He left us in no doubt about identity giving both the Latin and Irish names, identical to those listed by O'Sullivan Beare over a century previously, that is, *milvus* and pricane-na-cark (phonetic interpretation of *preachán na gcearc*).

Despite the fact that Thompson knew very little of the existence of the Red Kite in Ireland at that time (mid 19th century) there are a number of references from the end of the 18th to the beginning of the 19th century which indicate that it was still known locally then. The ordinance on game in the time of George III explicitly refers to kites as having a bounty of six pence on their heads. The fact that this was to take effect 'from and after the first day of July 1787' would indicate that kites were still sufficiently widespread in Ireland then as to constitute a nuisance. Thompson himself mentioned that he had been told by an old gamekeeper from Shane's Castle, Co. Antrim, that a few kites had been killed there (presumably around 1800), though no mention was made as to whether they were breeding. In the *Statistical survey of Londonderry* (Sampson, 1802) it states that 'milvus, kite, is frequently seen hovering over poultry', and Templeton, writing in his naturalist's report in the *Belfast Magazine* in 1808, states, 'next the kite proclaims another season – and time to shear your sheep'. The implication behind this last remark is that the kite was migratory, in contradiction to the remarks of Smith, from Cork. Could it have been that the Irish kite, as in Wales today, was only partially migratory, or perhaps a summer visitor to the north and an all-year-round resident in the south?

H. O'Sullivan recorded in his diaries (Callan, Kilkenny, 19 February 1828): 'I have seen this year only one Kite (*priacáin na gcearc*). There are thousands of crows for every one Kite, for it is good the crow does, or any grain-eating or worm-eating bird as they clean up the land of grubs, which would eat the seed and it is evil the Kites do, or any carnivorous bird which lives on pullets, young ducks or goslings, and, if carnivorous birds were numerous,

21 *Kite and Hawk? (14th-century)*

barn door fowl could not be reared and we would be without biped meat.' O'Sullivan was a keen observer of birds and no doubt he would have mentioned if the kite had been anything other than a rarity in his neighbourhood at the time. There is a reference to the kite as an uncommon resident in a list of birds known from Co. Londonderry (Kilrea district) in 1831.

A review of the decline of the Red Kite in Britain must echo its decline and disappearance in this country. Its former abundance in Britain is well documented. In 1571 it was an offence in London to kill kites due to their function in helping to clean up noisome refuse. James I protected them also because they were regarded as the best 'flights' for training falcons (see fig. 21). In fact this use of the kite as quarry for falcons persisted until the early years of the 19th century. Pennant (1777) wrote that kites were then widespread and abundant throughout England, though they were gradually disappearing from towns and cities as a result of improvements in drainage and refuse collection. In the country, obsessive ideas about the protection of game were beginning to adversely affect all birds of prey. Harting (1906) listed the many English counties in which the kite was common in the early 19th century. He attributed the decline to a combination of persecution and destruction of their woodland habitats, which became more

important as refuges as the birds grew scarcer. However, trapping and shooting by poultry farmers took a massive toll. The large estates with their stocks of small game – particularly Pheasants and Partridges – categorised kites as 'undesirable vermin' and were supported in their efforts at extermination by sympathetic legislation. During the 1680s and 1690s there are records of payments to coroners on the Isle of Man for avian 'vermin' killed in their parishes, mainly to protect stocks of Partridges. Specific mention is made of kites, Ravens, Scarcrows (Hooded Crows?) and Magpies. Kites figure particularly in returns from the parish of Lezayre, in the eastern part of the Isle (Girrad, 1978). The programme of extermination extended to Scotland where an inventory of birds of prey killed between the years of 1837 and 1840 listed 1795, including, unbelievably, 275 Red Kites (Perry, 1978).

Harting (1906) proposed that the cutting down of the woods and the felling and stubbing of trees was a factor in the decrease of the bird in England, and Thompson (1849) suggested that when Ireland was more richly wooded and less populous it was better circumstances for the kite. Paradoxically the authors of the *Atlas of breeding birds of Europe* point to the Red Kite being a scarce bird prior to the opening up of the primeval forests for low intensity cultivation. If this can be applied to Ireland, this raptor would

22 *Red Kite in flight*

have had its heyday in early medieval times when the (mainly secondary) woodland was being cut back to allow for the tens of thousands of ring fort farmsteads which were built at that time. It nevertheless appears to have been common and widespread up until the early 17th century at least. But the radical clearances of the 17th century appear to have eliminated, first the habitat (thus creating the circumstances whereby the bird could be attacked easily) and then the bird itself.

From the early 18th century onwards the gamekeepers of the mosaic of well-ordered estates carried out a concerted campaign of eradication of birds of prey, particularly large ones like Buzzards and kites. The gin trap and breech-loader were effective weapons but the kite was particularly susceptible to poison due to its propensity for carrion. Kites are known to roost communally, sometimes several dozen together. It would have been possible, therefore, for a knowledgeable gamekeeper to wipe out an entire population of kites in a single night.

The Red Kite disappeared from most of Britain by the mid 19th century, but a few pairs held on in south/central Wales spawning one of the most remarkable of modern conservation stories. The population is increasing slowly under strict protection, and breeding has recently spread beyond the Welsh border. In Ireland it is unlikely that the Red Kite survived as a breeding bird into the 19th century. Its decline and disappearance coincided with that of the Buzzard which was also widespread here until the latter 18th-century (Thompson et al.). Unlike the kite, the Buzzard managed to survive by adapting to cliff-nesting on the north-Antrim coast, particularly on Rathlin Island. The last decade has seen it revert to tree-nesting, increase in numbers and spread southwards once again: it now breeds in several counties in the Republic.

Given its success on the other side of the Irish sea, there is good reason to hope that the Red Kite will, in the not too distant future, come to nest in Ireland once again. In 1979 a pair was seen in Co. Wexford early in the year, but for one reason or another they did not remain to breed. This incident nevertheless showed that recolonisation without introduction is a real possibility. But Red Kites are still being eradicated. They are helpless against sheep carcasses baited to kill foxes. In the winter of 1990/91 two tagged birds were found dead (in Longford and Waterford) due to strychnine poisoning. Ironically, they were from a consignment of 19 brought over from Sweden as part of a recolonisation programme in Scotland. Another, apparently one of the same programme, was found poisoned in early 1997, at Cookstown, Co. Tyrone, have been seen earlier at Glenarm in Co. Antrim.

A sympathetic attitude from game concerns and farmers (who need to be informed of the cooperation achieved in the Scottish programme) is crucial if the inevitable reintroduction of the much-maligned Red Kite is to succeed.

NOTE

Names in Irish for the Kite (Red Kite)

Préachán, préchán, generic term for 'bird of prey' (commonly used nowadays for 'crow')

Préchán, glossing *milgus* (=*milvus*, Latin) Irish Glossary 507 (Early Irish)

Préachán ruaid (lit. 'red bird of prey') O'Grady Cat. 184.34

Priachán na ghearch – O'Sullivan Beare (*c*.1625)

Préacgán na gcearc, Préacan ceirteach – Bedel (1662)

Pricane-nae-cark (phonetic) – K'Eogh (1739)

Préacán na gcearch – O'Sullivan (1828)

Préachán ceartach, ceirteach, cirteach ('a kite') – Dinneen (1927)

Préachán ceirteach ('Red Kite' – *Scots Gaelic Dictionary*, Dwelly, 1901)

Cromán – Plunkett (1662), Joyce (1900), Ussher & Warren (1900)

Cromán cearc – Mhic Domhnaill (*c*.1850)

Cromán gobhlach (name for Kite in Scots Gaelic – Nicolaisen, 1963)

Clamhán gobhlach ('Kite, or salmon-tailed gled' – Dwelly)

Séig, name given to a bird of prey that may be taken (killed) without any share going to the landowner: the *Séig* may be treated thus because it 'carries off young pigs and hens'. Although the author identifies *Séig* with *Seabac* (hawk) the description best suits the Buzzard or perhaps the Red Kite. Old Irish Law-text (O'Davoren's Glossary, *per* Kelly, 1997).

Seirr shiach, described as 'the Kite of Cluain-Eo' by Donatus (in *Cellagh's death song*, 12th-century poem, trans. O'Grady).

Gabhlan gaoi (described as 'hawk' but means 'forked-tail of the wind': nowadays, the swift)

Egerton Glossary 10 (Middle Irish)

Eun or *Ean fionn*; *Cur* (*Cubar*); *Badhbh* are general terms for birds of prey. They are each translated as 'Kite' in a number of instances but could also be translated as 'Hooded Crow', or simply as 'Crow'. Other names have been used like *Clamhán* (and variations) but these are generally applied to the Buzzard rather than to the kite.

The question of the Irish name for the kite has been disputed by Lorcan O'Toole, who has carried out an independent investigation of the Irish material. It is his view a sustantial number of the references to the *préachán na gcearc* are to the Hen Harrier, for which, he believes, it is almost a literal translation (pers. comm., 1999)

THE EAGLES

The disappearance of the eagles represents the single most significant loss in our birdlife probably since the coming of man to Ireland. The loss is highly significant both in ecological terms (the predator/prey balance) and in overall, cultural terms. The eagle motif, for instance, one of the most unequivocal symbols of status and power, has left its mark in the literature and the arts down the ages, in Ireland as elsewhere. It is manifest today on the gateposts of some modern houses.

The extermination of the eagles has not, of course been restricted to Ireland or Britain. It has been going on throughout the world for as long as man has found his interests threatened by hooked beaks and talons.

23 *Gate-post eagles*

The two species which were resident in Ireland were the Golden Eagle and the White-tailed Eagle (also called the Sea Eagle). Research has shown that, for most of their period of coexistence with man, eagles were not differentiated as to species. Lockwood (1984) has stated that, in general, this did not occur until the 17th century. It is convenient, therefore, in investigating their early status to discuss them initially together even though to the ornithologist they are very different in appearance and habit.

There exist many long-standing names in Irish. Scharff (1918) lists eight: *Iolar; Fiolar/Fiolair; Iolrach; Acuil; Iolpa; Iolar ghréagach; Iolar tiomchiollach;* and there are a number of recent concoctions like *Iolar fírean* and *Iolar báneireaballach* for Golden and White-tailed Eagle, respectively. The oldest name is *Irar,* which has spawned *Ilar* and *Ilur* and the most frequent references are to *Iolar* and *Fiolar.*

Joyce, in his *Names of Irish Places* (1901), devotes a section to the role that the eagle has played in the naming of places throughout Ireland: 'In several wild and mountainous districts, formerly eagle haunts, they are remembered in local names … Iolar [Iller] is the common Irish word for an eagle and in anglicised forms it usually has the terminations -iller, -ilra, -ulra: as in Slieveanilra (the eagle's mountain, Co. Clare) and Coumaniller (the eagle's hollow) on the side of Keeper Hill, Co. Tipperary. The word assumes other forms, as for example in Drumillard, the name of four townlands in Co. Monaghan, which is the same as Drumiller in Co. Cavan (the ridge of the eagle). There is a hill on the borders of Tyrone and Derry called Craiganuller [and a rocky stack on the Antrim coast, Craigahullier] (the eagle's rock) … There are many high cliffs and mountainous districts which have been named for eagle's nests – *nadanuller.* They have in some cases given names to townlands.' Other corruptions have been suggested besides those listed by Joyce. Lough Ouler in Co. Wicklow, for instance, may be an -uller derivative (Fox).

The original Ordnance Survey maps and their associated notes specify the nest-sites of eagles. They are usually marked with the captions 'Eagle Rock', 'Eagle Mountain', 'Eagle Hill', 'Eagle Island', 'Eagle's Nest' and so on. Twelve are listed in the O.S.

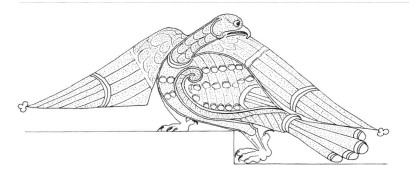

24 *Stepping eagle from the Book of Kells
(St John's Gospel), 8th-century*

Details but there are many more; on the map of the traditional Irish eagle eyries 37 are shown. It is not known now which species of eagle occupied these nests: those in the higher mountains were probably mainly Golden Eagles, while those along the coast were mainly White-tailed Eagles as were those on small coastal or lake islands.

There are references to eagles in the early Christian nature poetry. In the *Song of Manchin of Liath* (8th–10th-century) a comment is made about the 'lonesome scream' of the eagle. On the early Celtic crosses dating over an extended period, eagles are a favourite motif (from the 7th-century stone cross at Carndonagh, Donegal, to the 13th-century cross at Cashel, Tipperary). Eagles, which are unmistakably sea eagles from their profiles figure prominently also on the early carved slabs of Orkney.

The illuminated manuscripts of the 'Golden Age' of Celtic Art relied heavily on the stylised forms of eagles (as in the Book of Kells and the Book of Durrow) exhibiting not only familiarity but also, apparently, a certain affection. Love (1984) has shown that the eagles in the Scottish Celtic illuminations have characteristics of White-tailed and not Golden Eagles, and on inspection the same generally holds for those from early Irish work. The extra-large beak and bulky rear end to the body are characteristics of the former.

A number of the heraldic shields of old Irish families have the symbol of an eagle including those of the O'Raffertys, O'Moriartys, Lallys and O'Boylans. Birds other than eagles were common

heraldic symbols in the Middle Ages in Ireland. Hawks and crows (Choughs) and martins may be identified. Species which were considered to embody characteristics such as valour, dignity, independence etc. were chosen as suitable aggrandising symbols for the vainglorious clan chiefs.

In past times several commentators from Britain remarked about the abundance of eagles here. As mentioned earlier, Giraldus in the 12th century said that they were as numerous here as kites were in other countries. He alluded to the great number of birds of prey in general that were produced naturally in this country and, like many modern naturalists, wondered how their overall numbers remained static. Things have changed a lot in the last eight hundred years!

In the 14th century Ranulph Higden also commented on the abundance of eagles here. Stanihurst mentioned the breeding of eagles in Ireland in the 16th century, and Derricke in 1581 wrote: 'of feathered foules, there breeds the chief of all; a mightie foule, a goodlie birde, whom men doe eagle call. This builde her neast in highest toppe, of all the Oaken tree: Or in the craftiest place where of in Ireland many bee … ' O'Sullivan Beare (*c.*1625) was the first to differentiate between the eagles, describing them, their different appearance and behaviour. In the late 17th century, Gwithers, in a list of birds 'Common in Ireland, Rare in England' mentions the 'Sea Eagle' and in 'Birds found in Ireland, not in England' he mentions 'Eagle' (presumably Golden Eagle). This too is a suggestion of the relative abundance of both species in Ireland at that time, a relationship borne out in other later reports.

Eagles have been the subject of a considerable volume of folklore. Eagle tales from Ireland seem to be particularly colourful and, as is the way with folklore, often considerably removed from reality. The grasping eagle is often portrayed as a death-dealer, a pillager of the weak and vulnerable. While there may be some truth in the many stories about lambs, kids, cats and even dogs being lifted to their doom, those about humans babies being carried off (and invariably recovered by a hero figure) stretch beyond the limits of credibility. There are fables of eagles (presumably White-tailed Eagles) being pulled under by enormous fish, and others describing how they (Golden Eagles?) spooked deer into plunging from cliffs by grappling with their antlers. In Maxwell's *Wild sports of the West* (1843), there is a description of a battle between a sword-wielding lad (in a basket suspended over a precipitous cliff) and an eagle defending its nest. The lad survived the encounter, despite almost slashing the rope from which he was suspended. The eagles did so too, temporarily, at least.

One tale emanating from the mountainous district of Doolough in Co. Mayo, apparently in the early 17th century, concerns the misfortune of a local farmer, Roger O'Toole. Enraged about the death of a mare which he believed had been spooked to fall to its death by harassing eagles, O'Toole set out to wreak revenge on them. Having located the dead animal beneath the cliff, he removed the entrails and hid within the carcass. Eventually, when the eagles came to feed on the carrion, O'Toole sprung forth, grabbing the birds, each by the leg. To his horror, he was lifted from the ground by the birds and carried off in the direction of their eyrie. Aware that there would be no escape for him if he continued to hold on, he released his grip. Fatally injured, he was found by his sons (Berry, 1966).

While the story is embellished to the point of impossibility, it is regionally and temporally located and has resonances of authentic accounts of eagle-catching from Norway (see White-tailed Eagle).

The most bizarre tale, however, concerns the misfortune of a gamekeeper from Co. Meath. He came upon an eagle, sleeping, having gorged itself on the carcass of a dead sheep. Deciding to try and capture it, he threw his arms around its wings but the waking eagle responded by plunging a talon into his chest. In a battle which ensured, the gamekeeper succeeded in strangling the eagle. But unable to release himself from its excruciating grasp, he took out his knife and cut off the bird's leg. He then proceeded to go in search of medical attention with the eagle's talon still protruding from his chest! (Watters, 1853).

Colourful and occasionally humorous as the stories are, their unfair portrayal of eagles as ruthless predators have done nothing to improve their image among people at large.

□ White-tailed Eagle
△ Golden Eagle
○ Traditional Sites (Species unknown)
◑ Traditional Sites (probably White-tailed Eagle)
◮ Traditional Sites (probably Golden Eagle)

These are the locations of sites known to have been occupied by breeding eagles. Many were regular eyries and some, (those designated by ◑, ○, ◮, were traditional sites which may have been occupied for centuries. These traditional sites were denoted by either Irish or English names on old maps. Some of the traditional sites were undated until 1900.

25 *Eagles eyries (mainly early 19th-century)*

WHITE-TAILED (SEA) EAGLE

Haliaetus albicilla

This eagle has truly massive proportions, being noticeably bulkier and heavier than the sleeker Golden Eagle. The larger female has a wingspan often in excess of 8 feet (2.5 m). Otherwise the adults look alike and only they have a pure white tail. The plumage is overall grey-brown but older adults are mottled and often have a bleached-looking head. The young are dark all over but may be distinguished from the Golden by their bulk, their partially feathered legs and their flight profile, which is less pleasingly proportioned – more like a vulture than an eagle.

The White-tailed Eagle is in other respects vulturine. Carrion forms a major part of the diet, and groups gather around a carcass much in the fashion of vultures in the more arid parts of the world. Sea birds and especially fish are preyed upon, and in winter, when other food is scarce, they are known to scavenge in close proximity to man. They have been known to attack poultry.

Though there were some inland sites, the majority of Irish eyries were within sight of the sea on coastal cliffs and islets, usually in some inaccessible spot. In favoured localities several pairs might nest on a few miles of coast, in contrast to the solitary and highly territorial Golden Eagle. The nest is an untidy mass of sticks or other, similar available material. Both young are usually reared from the two eggs laid. (To date the introduced Scottish white-tails usually rear only one young eaglet: O'Toole, 1999.) The juveniles, which, for the summer, stay in close association with the adults, often wander widely during their first winter.

Formerly a widespread resident throughout Europe, the White-tailed Eagle has declined steadily and has now disappeared from almost all of western Europe, though it is holding its own in the east. It occupied ranges mainly in the north and west of Britain and Ireland in historical times. The population has been estimated at some 200 pairs (Love, 1983). This was based on an estimate of some 150 pairs in Scotland and about 50 in Ireland. In fact the pre-crash figure for Ireland was probably closer to 100 pairs (pers. research), suggesting a total population of about 250 pairs.

White-tailed Eagles, successfully re-introduced to the Isle of Rhum in Scotland (since 1983), are now breeding on the Hebrides and on the Scottish mainland.

Though generally non-migratory, there is considerable post-breeding dispersal of immatures, particularly those from the colder northern regions.

The White-tailed Eagle has a long and distinguished history as an Irish bird. Since Stelfox (in the early years of this century) identified its sketetal remains amid the piles of animal and bird bones recovered from the ancient crannog site at Lagore, Co. Meath, many other sites have come to light. These are widespread and range from the Mesolithic (Mount Sandel, near Coleraine), some 9000 years ago, to the medieval (Waterford), 13th- or 14th-century. The recovery sites are also widespread and include Lough Gur, Limerick (Neolithic) and Wood Quay, Dublin (Viking). At least five remains were uncovered at the latter site. It seems likely that many more will be uncovered.

At a number of these sites the remains of other birds of prey were also found, including the Buzzard and the Goshawk. It is curious how these normally elusive raptors could be found at human habitations. What they were used for, since there is no tradition of falconry in Ireland earlier than the Hiberno-Norman period, is not entirely clear. As in the case of the Red Kite, the White-tailed Eagle was more than likely a scavenger which, due to

Sea Eagle) Aquila albicilla Gue
Falco ossifragus?
I suppose this to be a young bird of
the first year not having the white ...

26 *White-tailed Eagle (immature) by John Templeton (c.1800)*

its large size, could be occasionally killed by a projectile and may have been used as a target. However, it has been speculated that raptors, trapped alive, may also have been kept as status symbols. Until about 1970 White-tailed Eagles were caught by hand in the Lofoten Islands off the coast of Norway. A man hiding in a stone cairn would gradually draw a piece of carrion on a string towards a small opening. A hidden accomplice would then grasp the eagle by the legs when within range. Many eagles were caught and killed in this way: the motivation was a demonstration of bravado (it took no small measure of that) or the retention of the wings and talons as trophies.

The eagle/fish motif is identifiable in the early literature. Eagles feeding on fish by taking them from the water – obvious references to White-tailed eagles – are mentioned in the Lives of the Saints from the first millennium. For instance: 'It [the eagle] brought fish from the river' – in the Lives of St Fintan and St Cuthbert. Several of the representations in the illuminated manuscripts of the 7th–10th-centuries depict eagles with fish (for example, the Book of Armagh). Though stylised and simplified, they look to be drawn from familiarity with the bird. There is no doubt that eagles were favourite themes, holding an important symbolic role in early Christian art. They can be traced from the book illuminations to the high crosses. The theme was obviously borrowed from the Celtic *La Tène*, where it figured prominently as the bird-head motif on many pre-Christian artefacts.

White-tailed Eagles often nest on small islands in lakes. They still do in Scandinavia and eastern Europe. There are quite a number of such sites, recorded on the first Ordnance Survey (1837–42) in the north and west of Ireland. Documentary evidence substantiates these nest-sites in a few instances. As stated earlier, the Golden Eagle was not generally a coastal breeder. Combining therefore, the coastal eyries with those known inland – including the lakeland sites – it has been possible to reconstruct the breeding range and approximate population of the eagles (more particularly the White-tailed) prior to the wholesale decline in the 19th century. The results for the White-tailed must be regarded as being more reliable than that for the Golden due to the fact that the latter occupied much larger territories and were more difficult to locate.

This reconstruction is revealing in that it shows the strong western bias in distribution, a pattern once reflected in Scotland and Iberia and still evident today in Iceland and Scandinavia. This pattern is likely to represent a relict distribution and not, as the archaeozoological evidence suggests, the former one prior to the decline. The historical evidence bears this out.

Giraldus' 12th-century remarks (in so far as they can be believed) suggest ubiquity, but O'Sullivan Beare, who identified this eagle (*Baulliar tieri*) by its white tail, referred to it as a 'western species' in the early 17th century. Also, he stated that it was found in 'both the country and the town' (Galway? Limerick?). Henry (1739) made mention of the abundance of these eagles in the north-west coast, referring specifically to Tory Island. Young, in his *Tour of Ireland 1776–79,* left no doubt as to species identity in this note from Mayo: 'They [eagles] also watch the salmon jumping and seize them even out of the water, by darting with that celerity of which they are such masters; this is so common that men with guns are out to kill and frighten them'.

Both Smith and Rutty provide us with sketchy references to Sea Eagles on the south and east coasts of Ireland and it does seem that there were a few traditional eyries here throughout the 18th century. It appears that there were eyries on Lambay Island (Dublin) and Bray Head (Wicklow) possibly as late as the beginning of the 19th century.

The distribution map which reflects the population in the early 19th century indicates the following: mainly western and coastal with concentrations in Donegal, Mayo, Galway and Kerry. Lesser numbers in Sligo, Leitrim and Cork; a few nesting pairs along the east Ulster coastline; a few pairs only in Waterford, Tipperary and Clare and eastern Dublin and Wicklow. This suggests a minimum of 75 sites attributable to the White-tailed Eagle: if these are combined with the eagle placenames such as 'Eagle rock' and Eagle island' it appears that there was of the order of 100 traditional eyries. Love's (1984) estimate of about 150 pairs (with adjustment, closer to 200 pairs) mainly in mainland Scotland and the western isles relates to the population before the early 19th-century decline. Ireland therefore held a significant fraction of the total population of the British Isles at that time.

In some places, particularly Donegal, nesting Sea Eagles remained locally common. Otway (1827) describes the scene at their Horn Head stronghold in romantic language: 'And here and there on some bolder and broader prominence too high from below and too deep from above to be accessible to man, were eagle's nests, and young ones as large as turkeys and the old birds from 30 to 40 at a time, floating in mid-air above, shrieking and challenging from on high our audacity in molesting their sovereignty.'

THE DECLINE OF THE EAGLES

Eagles were trapped and shot in Ireland for a long time before their status as breeding birds was threatened. There are references to the conflict long before the 19th century. With the refinement of game production and the establishment of reserves for that purpose professional gamekeepers started an all-out war against them and other large birds of prey. Shepherds and small farmers, concerned about the welfare of the flock or the coop, joined in a battle which for more than a century had been largely ineffectual but which eventually began to show results.

The literature is full of references to eagles attacking lambs. Most of it is probably exaggerated, but there is no denying that the occasional (usually sickly) lamb has fallen to eagles. This is well known to Scottish gamekeepers, mainly concerning Golden Eagles, but it is acknowledged that it constitutes only a small fraction of the overall prey. Wardens on grouse moors had more realistic grounds for complaint: the Red Grouse represented an important menu item for Golden Eagles. (Modern research in Scotland has disputed this.) White-tailed Eagles enjoy a wider diet with emphasis on fish and carrion, though they take sea birds when they are available. In winter, when seabirds are scarce, they may go for alternative food stocks such as poultry. It is for their eccentric rather than their normal behaviour that eagles have incurred the husbander's wrath.

'Whilst I was waiting in the last place [a farmyard in County Donegal] a large eagle flew directly over the yard and the fowl came running at us and making a noise from which I judged that they were alarmed by the noise of the eagle's wings' (Pococke, 1752). Young (1776–9) stated that 'they [eagles] do great mischief by carrying away lambs, poultry, etc.' and they were hunted on the north Antrim coast 'for carrying off lambs, turkeys, geese, of a tender age as well as ducks and hens'. In some areas eagles would brazenly harass fowl in the farmyards themselves.

Throughout the early 19th century eagles of both species but particularly the commoner and more approachable White-tailed Eagles were shot and trapped as the opportunities arose. Gradually the eagles became more wary, and only young birds remained naive enough to be caught in gin traps or allow a lethal shot to be discharged at them.

27 *Eagle and prey, panel above door, Church Street, Dingle (18th-century?)*

The introduction of strychnine as poison bait on carcasses in the mid 19th century heralded a radical change. Sir Ralph Payne-Gallway recorded the introduction of the poison and noted the subsequent rapid decline of the Sea Eagle in the west. Hoare, writing at this time, also stated that the White-tailed Eagles were so common there that the owner of Kylemore had recourse to the new poison. One gamekeeper poisoned eleven of these birds on Achill in a single year. The particularly cold winter of 1854 was thought to have had severe effect on the eagles by killing off game, their alternative food source at these times. Mayo shepherds, having been granted permission to raid eagles nests, did so with increased fervour in the interests of their flocks: an unquantifiable number of nests were destroyed in this way.

The campaign against the eagles gained momentum. In Donegal and Kerry there were reports of the kills. Thompson recorded the shooting of twelve or thirteen White-tails at Horn Head over a period of four years and others which had been killed by poison left primarily for foxes.

In certain places, however, the eagles survived despite the onslaught. On Clare Island, Co. Mayo, between 1867 and 1870 sea eagles contrived to take lambs and 'lifted geese from the cottiers' very doors due to no guns being allowed on the island'. West Mayo and the north Mayo coast continued to hold breeding White tailed-Eagles into the last quarter of the 19th century.

Golden Eagles being considerably scarcer and more wary were less affected by the campaign. They nevertheless declined steadily. Watters (1853) considered that at that time they were restricted to only the highest mountain ranges 'where they could not be persecuted by hunters and gamekeepers'. In the latter part of the 19th century they survived in the remotest parts of the mountains of Kerry, Connemara and in the Sligo/Leitrim/Donegal highlands.

As they became rarer, the eagles attained trophy status. They were sought after and paid for by the highest bidder, plenty of whom were around in Victorian times. Eagle eggs fetched high prices from avid oologists who were ready to pay 'danger money' to the daredevils who were prepared to scale cliffs to steal them. Eagle eggs bestowed on the collector a kind of status which others

28 *White-tailed Eagle with fish*

sought to achieve. Nor did the activity escape the attentions of those who should have known better. The eminent naturalist, A.G. More, writing to a friend in 1865 about a proposed fishing trip to Donegal, enquired, 'What will your honour wish me to pay for any eggs [eagle's eggs] I can pick up? What is your highest bid?' It is not known how many nests were robbed thus, but, happening when it did contemporaneous with advances in poisons and breech-loading firearms, it must have taken a toll of an already dwindling population.

When Ussher assessed the status of the White-tailed Eagle in 1894 he wrote: 'Still one or two pairs in Mayo and Kerry. Has been destroyed or driven from former breeding haunts in Donegal, Antrim, Down, Wicklow, Cork, Clare, Galway … ' When the status was again assessed in 1900 the White-tailed Eagle had become completely extinct in Ireland, the last nesting having occurred in 1898.

White-tailed Eagles continue to turn up in Ireland as rare visitors. Several, apparently stragglers from Scotland, have turned up in recent years. One was seen as recently as June 1998, off Inisheer island. On one famous occasion in September 1965 a party of five (almost certainly Sea Eagles) was seen off Cape Clear island, Co. Cork. The well-publicised pair* introduced on to Inisvickillaun, in the Blaskets (in 1992), refused to breed, and when one considers the scale of the effort involved in getting them to do so on Rhum, it is hardly surprising.

Nevertheless, the ongoing success of the Rhum project augurs well for the whole of its former range in Scotland and, given the proximity to the Ulster coast, it can surely only be a matter of time before successful breeding occurs here once again.

Love (1983) has made it clear that an overall strategy based as much on the maintenance of suitable habitat and abundant food supply, as an education and compensation programme, is a vital prerequisite to recolonisation. Perhaps we should be embarking upon this now.

---

* One of these is known to have died; the fate of the other is uncertain, though it may still be on the island (O'Toole, 1999).

GOLDEN EAGLE

Aquila chrysaetos

Often described as the most successful of the world's sixty species of eagle, the Golden Eagle is found widely throughout the world. Its main habitat is remote, mountainous terrain, though it is also found in forests. In Europe it is still found in the wilder tracts of many countries.

The female is very large, about 3 feet (0.9 m) long with a wing-span of some 7 feet (over 2 m). As in most raptors the male is slightly smaller. The plumage is dark brown with golden feathers on the nape giving the head a decidedly noble look. The legs are feathered, leaving only the bright yellow talons exposed. In the immature bird the tail is whitish at the base and there are large white wing flashes.

The Golden Eagle is a dynamic hunter, preying particularly on hares, rabbits and grouse in the mountains. However, a wide variety of animals and birds may be taken, and carrion is an item, though a less important one than live prey. Golden Eagles in Scotland have shown that they only rarely take lambs, and the evidence suggests that these are mainly sickly animals.

The Scottish nests and the majority of traditional Irish sites are on inland mountain crags, cliff-faces and corries, usually above the 2000 foot (600 m) contour, though occasionally considerably lower. In more wooded circumstances and times this eagle has been known to nest on tall trees. In the traditional sites the actual structure is a large bundle of sticks and heather which, through augmentation over the years, can assume enormous proportions.

Two eggs are laid often in early April but usually only one chick survives to adulthood. Being slow to mature and reluctant to fly, it usually stays in the nest for about three months. Of the two species of eagle, the Golden is the more retiring and territorial. Each pair may occupy a range extending up to 12,500 acres (5000 ha).

The Golden Eagle population in Britain was centred mainly in Scotland in historical times where there are more than 500 traditional eyries. The former population has been estimated at 650–700 pairs (Love, 1983). In 1986 it was estimated that there were about 420 breeding pairs in the highlands of Scotland. Since the 1940s the Golden Eagle has bred in a number of Scottish lowlands sites and has spread further south to occupy one or two in northern England.

It is mainly non-migratory, though immatures tend to disperse after the nesting season. It has been shown that only a small percentage of Scottish birds disperse to the south-west, towards Ireland.

Apart from the bones of a 'probable Golden Eagle' found in 13th- or 14th-century excavations at Lough Gur, no other fossils attributable to this species have been uncovered. The lack of tangible evidence is understandable in view of the species charac-teristic man-shunning behaviour.

A review of relative abundance of both eagles by the 19th-century naturalists substantiates the findings of the earlier com-mentators that the White-tailed was the commoner of the two. It is possible however that the sea eagle was simply seen more often than the shier Golden and was therefore assumed to be the more abundant. It is interesting that in Scotland there are four times as many traditional eyries for the Golden as for the White-tailed. Thompson stated that he never knew of the Golden Eagle nesting on coastal sites in Ireland and suggested that the eyries were restricted to the more mountainous areas. However, the most

recent nest site (in the 1950s) was on a sea cliff at Fair Head, so they may have occasionally used these locations. There are further historical hints that their territories may not have been as restricted as was formerly thought. It would seem that by the time the 19th-century naturalists carried out their county-by-county survey the Golden Eagle had already retreated considerably.

Those early writers who do specifically mention the Golden Eagle suggest that its haunts are the mountains of certain counties only. However, as the eyrie sites of the two species are known to overlap, the question of accuracy of identification arises. In the context of the poor or non-existent optical equipment of the time, this is a real consideration.

O'Flagherty mentions eagles spooking Red Deer in Connaught in the 17th century. Others (Molyneux et al.) noted similar behaviour, clearly suggestive of Golden rather than White-tailed Eagles. There are allusions to 'black eagles' which would also suggest the former rather than the latter species. The 18th-century writers give a fragmented account of the status of this bird. Smith mentions it from remote parts of Cork and Kerry; Henry (1739) notes Leitrim and Sligo; Sampson (1902), Donegal and other parts of Ulster. O'Donovan's survey (1830) mentions that the rock above Ballinascreen (in the Sperrins) was a traditional eyrie site. Generally, however, only snippets of imprecise information are forthcoming from these sources.

The map shows some fifty traditional Golden Eagle sites in Ireland. These are derived mainly from late 18th-and early 19th-century sources. It is interesting that research carried out by the late Jim Haine has come up with a very similar figure. While this number must in no way be considered the historical maximum, it may reasonably reflect the potential restored population: Haine listed 23 territories as the minimum number available for breeding Golden Eagles and those most likely to be colonised initially. (See the Note below.)

Although the decline of the Golden Eagle is well documented there is no need to look beyond the account of William Thompson who, obviously concerned about its imminent disappearance, made it his business to record it carefully. He traced back the status sometimes to the turn of the 18th and 19th century, county by county. By quoting the observations of contemporaries and extracts from county histories he was able to compile a reliable inventory. He stated (1849) that it was everywhere on the decline, 'becoming annually more rare' and was now 'very scarce in its former stronghold, County Kerry'. Watters (1853), too, outlined the range of this eagle and commented on its decline. The zoological journals of the later 19th century are replete with references to the decline and imminent extinction of the Golden Eagle throughout Ireland. An interesting detail about its declining status on the Dingle peninsula has emerged from Ussher's letters

29 *Golden Eagle: Historical Irish breeding records (J. Haine)*

(*per* J. Fox, *Irish Birds*, 1988). The correspondence is from a Dr Kane and is dated 29 March 1888:

> A few years since a pair of those birds [Golden Eagles] could be seen in every mountain range in this locality. In my shooting excursions I have often seen as many as twelve in a day. Owing as I believe to the frequent robbing of the nests and the trapping and laying of poison for vermin, this bird has now become nearly extinct. The last eagle I have seen was taken in this neighbourhood about eight years since with a large trap and chain hanging from his claw. I was pleased to hear that this year a pair of those birds had returned to their former breeding grounds. Their nests are made in a very high cliff overhanging a mountain lake (Lough Slot) in a mountain range between here [Annascaul] and Castlegregory. I have been informed by old inhabitants here that this cliff was always frequented by the eagles and that they bred there every year within their recollection which would go back nearly one hundred years …

Ussher, in a paper to the Royal Irish Academy in 1894, stated that 'Its chief breeding places are now in a few spots in west Mayo. It is breeding sparingly in west Donegal and probably in west Galway and Kerry. It has ceased to breed as formerly in Antrim, Tyrone, Down, Tipperary, Waterford, Leitrim and Sligo.' Six years later in *The birds of Ireland* (Ussher & Warren) it was mentioned as breeding only in Mayo and Donegal and only as a rarity. Having vanished from Donegal in 1912, a pair reportedly bred in a Mayo locality for a year or two later before they finally disappeared altogether. For many years afterwards single eagles were seen in suitable nesting locations along the north and west coasts but they were stragglers and were often shot.

A quite unprecedented sequence of nesting occurred from 1953 to 1959 at Fair Head, Co. Antrim, but these were eagles from Scotland and when breeding ceased in 1959 so did the real possibility of re-colonisation. However, in the knowledge that most Scottish eyries are presently re-occupied, it is certainly

30 *Golden Eagle by John Templeton, c.1800*

31 *Golden Eagle mobbed by Ravens*

possible that natural recolonisation could come about by the dispersal of young birds to Ulster – visible to the eagle's eye from Scotland's western highlands and islands. Work carried out by Jeff Watson, however, has shown that although dispersal of Scottish Golden eagles takes place in all directions it is noticeably light to the south-west, towards Ireland. In addition, the source from which wandering eagles might reasonably have come – the Kintyre peninsula, home to many pairs thirty years ago – is now almost without them due to afforestation which reduces their effective hunting territory. The potential for natural recolonisation to Ireland has thus been greatly reduced.

The case for positive reintroduction is probably stronger now than it has ever been. One or two wandering birds are seen most years, mostly in Co. Antrim but their appearance has been too sporadic to give rise to optimism about the possibility of nesting. A list of the rather sparse recent sightings, drawn up by the Irish Raptor Study Group, is not suggestive of imminent recolonisation and they are not optimistic about natural recolonisation.

NOTES

Historical Irish Golden Eagle Breeding Records by Jim Haine, IWC, February 1996

These records were compiled from the Ussher Bird Notes held in the Royal Irish Academy; Wm. Thompson, *Natural history of Ireland*, vol. 1 (1849); Ussher and Warren, *The Birds of Ireland* (1900) and *British Birds* nos. 47 & 55; grid reference and source(s) are given for each site recorded.

County Donegal, 12 Eyries
Muckish Mountain, B 990285, H. Saunders 1866
Glenveagh, B 985185, John McLennan 1905
Gweedore – Cronalaght, B 865250, Revd A.H. Delap 1878
Errigal Mountain, Dunlewy, B 930215, H.D. Barton 1891, Revd A.H. Delap 1891, John McLennan 1905
Rosheen Mountain, Dunfanaghy, C 008345, Wm. Thompson 1849
Horn Head, C 010425, James Rutherford 1887
Loughsalt Mountain, C 135265, John McLennan 1905
Lough Easke (Eske), G 995885, Revd A.H. Delap 1878
Lough Belshade, Blue Stack Mountains, G 980905, Revd A.H. Delap 1878
Slieve Tooey, G 630905, Revd A.H. Delap 1878
Lough Divina, Glencolumkille, G 575840, R Patterson 1900
Slieve League, G 565780, H.C. Hart 1890, W.J. Green 1897, J. Musgrave 1903

Counties Leitrim & Sligo, 7 Eyries
Skreen, Kings Mountain, G 705445, Col. Cooper 1891
Benbulben, G 690465, Col. Cooper 1891
Gleniff, G 758482, R. Tibirry 1872
Glencar Range, G 780445, Col. Cooper 1891, R. Warren 1892
Glenade, G825485, Gotottoenham 1880, M. Hook 1893
Cliffs of Aughris Head, G 505370, P.G. Symes 1891
Lough Talt, Ox Mountains, G 390170, H.W. Scroope 1907

County Mayo 14, Eyries
Croaghan Mountain, Achill, F 560060, J.R. Sheridan 1891
Slievemore, Achill, F 655 090, J.R. Sheridan 1891
Curraum Mountain, Achill, L 755965, J.S. Palmer 1884, G.F. Ormerod 1892, John Le Warren Page 1909/10
Knockmore, Clare Island, L 675860 *Irish Naturalists Journal* 1891
Portacloy, F 843860, W.M.J. Doherty (NT) 1907
Porturlin, F885430, N.O. Brien Hickson 1912
Benwee Head, F 815450, W.M.J. Doherty (NT) 1907
Mount Eagle, Nephin Beg Range, G 005045, R. Warren 1911

Corslieve, Nephin Beg Range, F 918125, H. Scroope 1912
Burishoole, Bengorm, F 930002, John Le Warren Page 1911
Dhuloye, Mweelrea, *Irsh Naturalist* 1898
Delphi, Mweelrea, L835665, R. Warren 1892
Errif Mountains, Maumtrasna, L 965665, John Hearne 1887, G.F. Ormerod 1892
Erris Valley, Maumkelly, F. 915172, R. Warren 1892

County Galway, 2 Eyries
Twelve Pins, L 780520, W.M. Thompson 1849
Maumturk Range, L 890535, McKeown 1888, W.H. Good 1890

Counties Tipperary & Waterford, 4 Eyries
Coumshingaun Lough, Comeragh Mts, S 320115, Rev P.A. Keonig 1868
Bay Lough, Knockmealdown Mountains, S 003115, W.M. Corbert 1882
Galtee Mountains, R 895215 Revd R.M. Milne 1892
Slieve Kimalta, Keeper Hill, R 870665 W.M. Corbert 1882

Counties Kerry & Cork, 13 Eyries
Upper Lake, Glena, V 925840, M. Moriarty 1887
Caher Mountains, MacGillycuddy Reeks, V 784844, W.M. Ashton-Hackett 1878

Mount Brandon, Dingle Pennisula, Q 465118, A.R. Brooke 1870, M. Moriarty 1887
Cominard, Lough Slat, Dingle Pennisula, Q 610075, M. Moriarty 1887
Glenbeigh, V 649838, J. Chisholm 1891, C.V. Stoney 1900
Coumasausin Lakes, V 757905, T.W. McCormick 1896
Cloon Lough, V 710775 J. Breen 1890, G. Fevees 1898
Lough Inchiquin, Caha Mountains, V 855625, M. McCormick 1896, G.A. Brennan 1898
Glen Rastel, Caha Mountains, V 820573, G.A. Brennan 1898
Glenmore Lough, Caha Mountains, V 760540, M. McCormick 1896, G.A. Brennan 1898
Glen Lough, Ardrigue River, Caha Mountains, V 848535, G.A .Brennan 1898
Barley Lake, Caha Mountains, V 890550, Rev. Pike 1893, G.A. Brennan 1898
Derrynasaggart Mountains, W 170830, Revd N. Hemyns 1887

Counties Fermanagh, Tyrone, Antrim & Down, 5 Eyries
Aghamore, South Fermanagh, H 105450, T. Wilson 1873
Sawel Mountain, Sperrins, H 620960, H. Saunders 1865, Rev C. Irvine 1894
Fair Head, D 190440, G.D. Deane 1962
Glenariff, Cushendall, D 220260, H.D. Barton 1894
Eagles Rock, Mourne Mountains, J 350255, A. Bennington 1954 (last bred 1865)

MARSH HARRIER

Circus aeruginosus

This fine bird of prey, the largest of the harriers, is almost the size of the Red Kite. The female, as in other raptors is larger than the male: it is nearly 2 feet (0.55 m) long and has a wingspan of over 4 feet (1.25 m). The long-winged, long-tailed profile is similar to that of the kite, though the tail is square-ended, not forked.

The flight of the Marsh Harrier is especially distinctive: it glides a lot, usually close above the ground with the wings held slightly above the horizontal. It can thus be identified, even from a considerable distance across a wetland. On sighting prey it frequently hovers before pouncing. Occasionally, in sunny weather or on migration it will soar.

The primary habitat is freshwater wetland, especially reed swamp. The territory extends to marginal land adjacent to bogs fens and marshes and to adjacent agricultural land. The range of a pair may be 1250 acres (500 ha) of which a quarter, at least, is normally reedbed. Prey consists mainly of aquatic birds like Moorhens, Coots, Water Rails, various ducks and their young and waders such as Snipe. Small mammals are also taken.

The nest, a large platform of sticks and reeds, is normally well-hidden in a reedbed, and four or five eggs, laid in May, is the normal clutch.

The female Marsh Harrier, besides being larger than the male, is also more striking in appearance. The plumage is chocolate brown (black at a distance) with a creamy top to the head, throat and front edge of the wings. The male is reddish-brown (paler on the head) and streaked darker on the underparts. Large patches on the wings and tail are ash-grey.

A widespread resident throughout much of central and western Europe, the Marsh Harrier is also a partial migrant. Many winter in Mediterranean countries.

Marsh Harrier bones have been found at a number of archaeological sites, in various parts of Ireland, covering a wide span of time. Those from Lough Gur (though not dated) were undoubtedly prehistoric. They were found among an assemblage of bones of the wetland avifauna consisting of ducks, geese, swans, gulls and others. During the excavations at Ballinderry crannog, Co. Westmeath, large numbers of bird bones were found which were also mainly those of waterfowl. In addition there were several which belonged to a larger bird of prey. Henken, who examined and classified the bones, thought that these were the skeletal remains of Marsh Harriers: they dated from the Early Christian period. The remains of at least two Marsh Harriers were uncovered from 10th–11th-century excavations at Fishamble Street, Dublin. These were found along with the remains of more than two dozen other birds (waterfowl and waders), most of which were obviously eaten. But the list included seven other kinds of birds of prey – Hen Harriers, Peregrine Falcons, Sparrow Hawk, eagles, kites, Ospreys and no less than ten Buzzards (T. O'Sullivan).

What were all these birds of prey doing at such a populated place? Were they caught and eaten along with the wildfowl? Could they have been kept as pets or for hunting? Were they simply dead scavengers? It would be reasonable to think of the kites and perhaps the White-tailed Eagles as scavengers, killed perhaps as targets. But the Harriers? Could it be that in former times this normally shier than shy bird adopted a different lifestyle – joining the other scavengers for easy pickings – or could it have been kept as a pet or a living talisman by status-conscious Vikings? Could it be (though this seems most improbable) that birds of prey, including Marsh Harriers, were imports from other countries – Scandinavia perhaps?

It is to be hoped that with the ongoing programme of excavations from other early urban centres the answers to these and other questions will emerge.

The traditional Irish name for the Marsh Harrier is less forthcoming than the names of other raptors. Why this should be is difficult to fathom: after all, it was the commonest large Irish raptor (according to Watters) in the early 19th century; it must therefore have had a well-known Irish name. It may have been variously described along with other Buzzard-like birds under a single generic name like *Éan fionn, Cur, Préachán, Cromán, Séig*. But, since there are specific traditional names for the distinctive Red Kite (*Préachán na gcearc*) and Buzzard (*Clamhán*) there should, in view of its singular appearance and habitat preference, also be one for the Marsh Harrier. That listed by Moriarty (1967), *Cromán mona*, would seem to be the most likely candidate. Significantly, Nicolaisen (1963) lists *Cromán mona* in Scots Gaelic. This is certainly the most appropriate name given the boggy habitat favoured and that 'Moor buzzard' is the traditional English name for the bird (Lockwood, 1984).

The names commonly applied in modern lists of Irish birds are *Cromán* and *Préachán na gcearc* but as already mentioned these (or variations) were also listed as ancient names for the Buzzard and the Red Kite. Thompson (1849) said that he knew of the Marsh Harrier being called 'kite' in the south of Ireland, and it is quite possible that, with the earlier disappearance of the Red Kite, the name was simply transferred. The names 'kite' and 'gled' (derived from the gliding flight) were also applied to the Hen Harrier, in parts of Ulster (Braidwood, Deane et al.).

In an apparent allusion to the haunt of a Marsh Harrier a wetland in Co. Kerry is named Curraghatouk. This, according to Joyce (1901) is a corruption of *Corrach an tSheabhaic*, which means literally the 'marsh of the hawk'. So this placename may introduce yet another generic name into the story. In Plunkett's Irish/Latin dictionary (1662) under the Latin name *Circus* (the harrier family) is the Irish *Fos gné sheibhce*, a 'type of hawk', indicating that, although it was specifically known, there was apparently no standard Irish name known to the author. Rutty (1772) made the first positive and reliable mention of the Marsh Harrier, calling it the 'Moor buzzard' and stating that it had been shot in Co. Dublin. Unfortunately he gives no Irish name nor clue to the bird's distribution or abundance.

Thompson in the mid 19th century considered the Marsh Harrier to be 'a widespread nester moving to low ground outside the breeding season' and listed three northern counties in which he knew it to be resident. Watters, Thompson's contemporary, considered it to be 'the commonest large raptor in Ireland' at this time. Given that at that stage the Red Kite and the Goshawk had already gone and that the Buzzard and the eagles had been much reduced, there can be little doubt as to the accuracy of his comment. The references cite Antrim, Down, Derry, Dublin, Kerry, Cork, Fermanagh, Tipperary, Donegal, Leitrim as definite breeding counties for this harrier until at least the mid 19th century. In some places they were even described as being common. Locations such as Lough Iron, Co. Westmeath, the bogs of Co. Offaly, the fens of Co. Laois, the Lough Erne complex and the lakes of Co. Galway were particular strongholds.

Like the Bittern, the Marsh Harrier is a habitat-specific bird; undisturbed, extensive wetland is its bailiwick. Unlike the Bittern, however, it can utilise agricultural and other marginal and reclaimed land peripheral to the wetland as subsidiary habitat. Extensive habitats of this kind were commonplace – in raised bog and cutaway fringes and in fens, until the mid 19th century. The Bog Commissioner's maps of 1810–14 make it clear that, apart from some superficial drainage adjacent to demesnes and the peripheral cut-away of turbary, the wetlands were largely unaltered then.

The first O.S. maps, to a great extent, corroborate the original survey. However, the reclamation brought in train by the

On April [1893] I visited the extensive marshes of Lord Castletown, near Granston Manor, In Queen's County [Laois]. These cover more than 800 statute acres, and are a paradise for ducks, lapwings, redshanks, coots, and gulls. High over these marshes I saw sweeping in curves, three marsh harriers, two together and a third apart. After two or three flaps, they would sail around, holding their wings slanting upwards. In wind, however, they are seen flying low, almost beating the tops of the flags with their wings in quest of their prey. As they are seen there in all seasons they must breed.

The evidence relating to the Marsh Harrier's decline refers repeatedly to their having been shot by gamekeepers: this would have been the second major factor in their disappearance. Why they should have been a target for gamekeeper's guns remains something of a mystery since Marsh Harriers take mostly aquatic prey. Watters clarifies, however: 'occasionally even the Teal or the Mallard are surprised and struck down, for which reason in some parts of the country they are destroyed by the gamekeepers'; also: 'five and seven birds [Marsh Harriers] have been trapped at the same time near the lake of Ballynacargy and on the estate of Sir W. Levinge in Westmeath'. Ussher wrote, 'Of all the birds of prey the harriers are easiest to kill, from their low, deliberate flight.' Even naturalists took their toll. In the summer of 1850 A.G. More and W. Shawe-Taylor, better known for their collecting of plant and insect specimens, also shot and collected various birds of prey – including Marsh Harriers – at Castle Taylor in south Galway (Nelson, 1991).

John Feehan, writing about the disappearance of the Marsh Harrier from Co. Laois, suggested that drainage, shooting and egg-collecting were the main causes. The decline was noted by Ussher & Warren at the turn of the century. Ussher stated that the vibrant population 'preserved' by Lord Castletown on his Co. Laois estate was a 'pleasant contrast to the monotonous tale of extermination from other places'. They stated that it was practically wiped out by gamekeepers in counties Westmeath and Offaly though it still bred there in 1889. Galway was also mentioned as a stronghold county in the later 19th century.

32 *Marsh Harrier breeding distribution (counties), 19th century*

Commissioners of Public Works was such that over 250,000 acres (100,000 ha) of wetland was drained or alleviated from prolonged flooding under the 1842 Act. These works continued for more than a decade and under the 1863 revision were continued for a further period. The effect on wetland-dependent birds must have been devastating. The disturbance caused by large gangs of men (hundreds together; tens of thousands in all throughout the duration of the works) would of itself have been significant. The fact that the drainage was carried out mainly during the summer (when the water-table was at its lowest) must also have put intolerable pressure on birds like breeding harriers.

Ussher provides us with a vivid description of Marsh Harriers in a wetland that managed to avoid destruction as a result of the first period of drainage.

33 *Marsh Harrier (female or immature) feeding on Water Rail, by Dunscombe Parker, c.1840*

The 'great lake' enigmatically referred to by Ussher at which he saw the Marsh Harrier in June 1897 and at which he had been informed it continued to breed was surely Lough Corrib. To this day extensive fenland, good harrier habitat, remains along its eastern fringe. The majority of sightings in the late 19th century, however, were of unattached birds and they were recorded from a wide range of localities. The *Belfast Guide* (1874) spoke of this species as 'being not very uncommon' on the Ulster moors. But Ussher and Warren (1900) stated that it was then no more than a straggler. Isolated harriers were recorded from the Antrim plateau – where Brunton (1871) reported that they had been exterminated years previously – and in the Donegal/Leitrim area (Allingham,

1879). (There is the possibility of mistaken identity in regard to harriers on the Antrim plateau: this upland heath plateau with little true wetland would be more suitable to the Hen Harrier.)

British ornithologists, acutely aware of a parallel decline in Britain, knew of its status in Ireland. In 1889 Saunders wrote: 'Ireland offers many more congenial situations, and the bird was formerly common.' Since 1840 the keepers have nearly succeeded in exterminating it by the use of poison. Coward in his *Birds of the British Isles and their eggs* (1920) thought that 'a few pairs still nest, or attempt to nest, in Norfolk and in the wildest bogs of Ireland.'

The last Irish strongholds were the fens and bogs of Co. Laois. Lord Castletown told Ussher (in 1896) that he had seen five Marsh

34 *Marsh Harrier in flight*

Harriers on the wing at once, in the vast curragh marshes near Durrow, and that he believed that his estates (which included these marshes) provided safe breeding sanctuary to about eight pairs. Ruttledge was of the opinion that they continued to nest there into the 20th century and may have continued until about 1917. By that time the Marsh Harrier had ceased to breed elsewhere in Ireland.

Unfortunately and despite the protestations of Lord Castletown, the Curragh marshes were substantially drained in the early years of the new state. The disturbance caused (part of the old bridge in Durrow was blown up to facilitate the drainage) and the subsequent radical transformation of the terrain rendered it more or less unsuitable for the possibility of return.

The further episode of wetland drainage under the 1945 Arterial Drainage Act not only eliminated many tracts of reed swamp that survived the 19th-century drainage, but reduced to insignificant fragments many wetlands that had remained (or regrown) to a size that would have been immediately attractive to harriers.

Nowadays, though it has been seen on an regular basis in recent years, the Marsh Harrier is still a rare visitor. It has been seen in spring in suitable breeding habitat, particularly along the south coast and a pair has summered in Co. Wexford. A pair was seen in May 1995 at Ballycotton, Co. Cork in suitable breeding habitat. They did not, however, breed.

> The on-high skydance of the male Marsh Harrier, diving, rolling, looping and screaming in his March or April trance, is some performance, and one to be coveted indeed. Now is the right stage to implement any additional conservation measures for reedbed sites with adjacent foraging potential, including bringing under control any disturbance to such sites in spring and summer.
>
> (D. Scott, *The Marsh Harrier in Northern Ireland*, 1995)

In view of the tentative recovery which it has made in England in recent decades, its return to Ireland must also be considered a real possibility. A few sufficiently large tracts of suitable habitat remain which could support breeding harriers but, apart from Pollardstown fen in Co. Kildare and a few others (mainly in the south of the country) which enjoy national heritage status, their future is by no means assured. Given the ease with which Marsh Harriers may be shot by renegade wildfowlers, the question of its successful return and survival remains doubtful. However, since this raptor does not conflict with farming interests and it is no longer being sought after for collections, it might ultimately be easier to protect than other large raptors.

NOTE

Stuffed specimens of Marsh Harriers are to be found in private collections and in the museums. Many of these passed through the hands of the taxidermist firms which in the 19th and early 20th centuries had thriving businesses in the main cities. The inventory list of Williams & Son, Dublin-based taxidermists, indicate an average of one or two annually up to 1916 but very few in the ensuing years. Only three, all from Ulster, were listed by the Belfast-based Sheals Bros. However, two were recorded by Ussher and Warren as having been sent for preservation to Mr Sheals in August 1884 by Mr Macpherson, a gamekeeper from Co. Leitrim.

OSPREY

Pandion haliaetus

The Osprey – the famous 'fish-hawk' of many countries – is regarded as one of the classical birds of prey. Due to its large size (equivalent to the Red Kite or Marsh Harrier), distinctive markings and singular fishing behaviour, it has consistently attracted attention down the ages. In flight it looks particularly large due to its 5 feet (1.6 m) wingspan. The Osprey soars and glides less than other raptors but uses its long flexible wings to hover over water, then dive feet-first for fish. It feeds exclusively on fish, taking both coarse fish and salmonids in lakes and rivers and inshore marine fish (mullet etc.) in estuaries and other coastal habitats. Fish farms are occasionally visited by opportunistic Ospreys.

The Osprey is strikingly patterned. Its upper parts, including a mask-like eye stripe, are dark brown. The head (which has a shaggy, crested nape) and the underparts are white except for a brownish smudge on the breast. The sexes look alike but can be separated in the field: the female is more distinctly marked on the underparts. In the air the Osprey looks quite unlike any other raptor with its contrasting plumage – predominantly pale below, dark above.

Though its range is virtually world-wide, European Ospreys are found mainly in the east and north. In some locations they breed along the coast and even sometimes colonially, but in Scotland they are solitary, favouring inland sites on lochs or large rivers. The nest is usually at the top of a very high tree (often an old pine) and when occupied year after year can develop into an enormous structure. In the past ruined castles on loch islands or promontories were used as alternative sites to tree tops. In general two or three eggs are laid in April and the young are ready to fly by July or August.

Ospreys are highly migratory. The majority winter in Equatorial Africa. Those from western Europe over-winter in the fish-filled estuaries and mangrove creeks of West Africa.

It is highly fortuitous that fossil evidence has been found here. The Osprey would be much less likely to come into close contact with man and his activities than, for instance, the scavenging Red Kite or the Sea Eagle. Osprey bones, dating from the 10th /11th-century, were unearthed in excavations at Fishamble Street (Dublin) within the past few years (T. O'Sullivan). The remains of two birds were found showing that it was hardly a random event that brought them into fatal contact with humans. Since Ospreys eat only fish that they catch themselves, the scavenging option can be discarded (despite the temptation implicit in the name of the site). Assuming that they were inhabitants of the Liffey in medieval times, could they have fallen victim to Viking market wardens concerned about pressures on local fish stocks? There is a medieval reference to Ospreys being attacked at monastic fish pools in England (*per* O'Toole). It could be that they constituted more of a nuisance in the past than we can envisage now.

Aside from the scarce archaeozoological material, the Osprey's claim as a former Irish bird can be established through the Irish language. There are also a few first-hand historical accounts. As with other distinctive birds like the Capercaillie, Bittern and Crane, there is also an element of folklore which has helped in the inquiry.

Moriarty & Fox (1997) have stated that the fanciful illustration of the eagle of St John in the Book of Armagh (9th-century) looks more like an Osprey. It is certainly more like an Osprey in its

35 *Osprey head and tonsured head of a monk*

proportions and seems to be hovering as would an Osprey, with a fish in its talons. The White-tailed Eagle is the only other bird that it could possibly represent, but the resemblance is tenuous, to say the least. However, as the eagle (unspecified) is the traditional symbol of St John the Evangelist, perhaps the monastic illustrator was taking even more licence than was customary in depicting a bird with which he was familiar.

There is an early literary reference (in *Scél Tuain meic Cairill*, an Irish saga, also from about the 9th century) to a bird of prey, *Séig*, which caught salmon in its talons. The bird intended must surely have been an Osprey (see p. 55).

The usual name, however, is *Iascaire cairneach* (lit., tonsured fisherman), a most apposite name emphasising as it does the bird's clearly demarcated white crown. The name must date from a time when tonsured people – monks – were commonplace, as in the early Christian church. Furthermore, this anthropomorphism suggests an empathy, even affection, for the bird by those who were familiar with it.

Scharff (1915) lists *Cóirneach* along with a name *Gríbh* which is to be found in Bedel's 17th-century translation of the Bible and refers to another (indeterminable) species. Another listed name *Préachán Ceannan* is similar to that mentioned by O'Sullivan Beare, *Iascaire Ceannann*, which means literally [white] headed

bird of prey/fisherman and as such is similar to *Cairneach* or *Cóirneach*. In the dictionary of modern Irish words *Ospróg* is given but it cannot be located in Early or Middle Irish sources so it is probably of recent origin. A Scottish name given by Nicolaisen, *Iolaire Uisge* (lit., 'water eagle') is identical to a name listed by Scharff but this is surely the Sea eagle. Interestingly, the name of a Scottish loch with a traditional Osprey breeding site is Loch-na-Claise Carnaich (Gooders, 1983) and it would follow that, just as in Ireland, the Osprey was in earlier times called by this name in Scotland.

The earliest description of the Osprey in Ireland comes from the pen of Giraldus: 'They hover quietly on their wings high up in the air over the waves of the sea. In this way they can see more clearly down into the depths below. Then seeing with their sharp eyes through such a great distance of air and troubled water little fishes hiding below the waves, they dive down with amazing speed'. This is a clear description of the fishing behaviour of the Osprey. Fox, in consultation with the Irish scholar Fergus Kelly (who looked at the original Latin text) extracted other snippets from Giraldus, including its size ('smaller than an eagle, bigger than a hawk') and that it was 'numerous' here. Giraldus has received a fair amount of criticism by reviewers who hanker for verification of the Welshman's more implausible claims. While this is not usually forthcoming in contemporary sources, it may become available centuries later, ecological conditions having changed little throughout the Middle Ages. Thus, a description of the Osprey in a manuscript attributed to Molyneux (1684) can be read both as a clear commentary on its former existence here and as a vindication of Giraldus' remarks:

An Osprey or Water Eagle is a large bird of prey – lesser than an eagle and bigger than a falcon, it builds in the ruins of some old stone works near great Loughs and sometimes (but very seldom) in old big trees. The feet are both like the feet of an Eagle. I have [seen] several of their Eyeries [*sic*], and they constantly have three old ones to each Eyerie and the Natives of this Kingdom [Ireland] say this never fails; I fancy it is an odd young one of the preceding year that does not

couple but on better consideration this is hardly probable; This fowles way of preying is She hovers a great height over a deep water like a Castrell [Kestrel], and of a sudden comes down with great force and going oft quite under water rises with a large fish in her talons which Shee seldom misses: this greatly amased [*sic*] me and I was the more inquisitive because I never saw this bird in England, Wales or Scotland: they report a great many ridiculous things [about this] bird, as that it drops fat out of its rump into the water which attracts and stuns the fish: that it has one foot like a goose, the other like a hawk with [which] it seases [*sic*] its prey and swims with the other, but this is apparently false. I one [once] discoursed with an old Irish man that had lived all his Days where they used to fish and he assured me that they hover thus when the sun shines, and the fish, seeing their Shadow on the ground at the bottom of the water, fly from the motion of the Shadow to the top of the water where they are taken ...

Even without a plumage description it is obvious that Molyneux is describing a first-hand encounter with an Osprey. The myth of the

different feet was first described by Giraldus. It is easy to see how this and others (like the dropping of the fat from the rump to attract fish) could have been kept alive in folklore, in the absence of modern science, by people who loved storytelling.

We don't know to where the incident refers, but in another passage dated *c.*1709, in a description of Co. Leitrim, Molyneux states that 'there are several eyerys [*sic*] of Eagles and also of Ospreys. These latter build usually on old walls near great rivers or loughes and feed on fish yet the Common tradition that they have one foot like a Goose and the other like a Hawke is daily proved to be a vulgar error ...' This is an unambiguous reference to breeding Ospreys (and probably also White-tailed Eagles) in Co. Leitrim at the beginning of the 18th century. This is corroborated in a letter dated 23 April 1683 which states as much (was it also from Molyneux?). In another dated 24 July 1790 from Robert Marsham to Gilbert White (author of the *Natural history of Selborne*) the writer states that when he was on the Lake of Killarney he was told that a pair of Ospreys that 'Yearly nested on an island of rock in that lake used to drive off their young as soon as they are able to provide for themselves' (from *The life and letters of Gilbert White,* 1922). This is interesting in view of the fact that Goldsmith knew of breeding 'Eagles' on Lough Leane, Killarney, in the latter 18th century. It seems likely that Goldsmith was going on second-hand information, for he certainly knew his birds, and more than likely his 'eagles' were the resident pair of Ospreys. The original O.S. map actually specifies the rock/island and it is still called 'Osprey Island' today. The island lies within gunshot of the shore and leaves one wondering about breeding success and the eventual disappearance of the birds from the site.

On his visit to Lough Key, Co. Roscommon, on his 1779 Connaught tour, Beranger remarked that 'There is an Eagle's or an Osprey's nest on top [of a ruined castle on an island] and we could hear distinctly the cry of young ones'. This is surely also a reference to breeding Ospreys rather than White-tailed Eagles.

Strangely, the 18th-century natural history writers make no mention of the Osprey as one of Ireland's breeding birds, formerly or otherwise, and even the 19th-century writers speak of it only as

36 *Osprey and fish: carved stone, Tuam (late 18th/early 19th century)*

a rare visitor. Thompson (1849) does mention summer sightings from the Lakes of Killarney and from the vicinity of Lough Corrib, Co. Galway. Ussher & Warren (1900) list numerous occurrences of the Osprey in Ireland but as visitors only. They do allude to Giraldus' comments, however, and suggest that the Osprey must have been commoner then. There may be other references in the archives which may eventually come to light. However, these would be important primarily in terms of establishing favoured haunts, since the Osprey's claim as a former Irish breeding bird has already been established.

O'Sullivan's knowledge of it (in West Cork or Kerry), the evidence from Leitrim, from Roscommon and perhaps from Galway point to it having a distinctly western bias in its former distribution. This was probably as much to do with remoteness and lower levels of disturbance as with the fact that the concentration of good fish lakes is west of the Shannon.

The well-documented Scottish experience may also provide us with background to the decline of the Osprey here. St John recorded that Ospreys were widespread breeding birds in the lochs of Scotland in the early 19th century but were slaughtered by gamekeepers who regarded them as a serious threat to fish-stocks. In the period 1837 to 1840 at least 18 Ospreys were wiped out in the Glen Garry area alone (Bijleveld, 1974). Throughout the 19th century the birds themselves and their eggs became fashionable trophies for British collectors. More (1865) remarked that the Osprey and its eggs were so sought after that only a few pairs managed to breed undisturbed in Scotland. Between 1843 and 1899, Ospreys nested 24 times at Loch-an-Eilein and were robbed 15 times (Ritchie, 1920). Two brothers, Lewis and William Dunbar, are recorded as having 'collected' the last breeding Scottish Ospreys. They nested on sporadically, however, until eventually ceasing completely in the early years of the 20th century.

In Ireland the 'trophy craze' was not as intense and occurred later than in Britain. So while Ospreys shot in Ireland can be found in museums and collections, they are usually 'one-offs', rare stragglers from Scandinavia or elsewhere, shot on migration and usually from the second half of the 19th century or later. The disappearance of native Ospreys took place earlier – prior to the time of Thompson and Templeton (both astute observers) in the first part of the 19th-century. It may, in fact, have been as early as the mid 18th century, since Rutty and Smith were quiet on the subject. It is likely to have happened therefore as a consequence of gamekeeper zeal, in the interests of maintaining the fish stocks of the large estates on which the Ospreys bred. The knowledge that we have now, that Ospreys take mainly weak or sickly fish was not available at the time, and the fact that sizeable salmon and trout were being removed from the lake or river waters under the noses of the owners was sufficient justification to attack them with the means that were available to the game-keepers – the flintlock fowling piece and the baited gin trap. In effect, the Osprey was a broad-daylight poacher – an aerial counterpart of the despicable peasant who, with bog-deal torch and under the cover of darkness, carried on his own surreptitious enterprise. Young, on his tour of the west of Ireland (1776–9), mentioned that firearms were used against the White-tailed Eagle in Mayo by salmon-fishermen due to competition for salmon stocks. No doubt the same persecution was aimed at the Osprey. An entry in John Templeton's diary for 16 March 1821 reads: 'Saw a fine specimen of Falco ossifragus [archaic name for the Osprey] caught at Tempo, in the Domain of Mr Tennant in the County Fermanagh. When taken it had a trap and chain weighing above six pounds attached to its leg, the chain of which trap had gotten entangled on a tree.'

It is heartening to trace the recovery made by the Osprey to Scotland as a returned breeding bird, since the first tentative success in 1954. The Loch Garten birds became as revered as royalty and soon their offspring began to occupy suitable habitat elsewhere. By 1974 there were 15 pairs nesting and under the unqualified protection (mainly voluntary) afforded them, they continued to prosper. Presently the population hovers around a healthy 100 pairs. There is no doubt that, with the high level of public interest, countering whatever opposition that they might receive from fishing interests, the Ospreys are back for good this time.

37 *Osprey overhead*

Ornithological research has linked the recolonisation to recent climate change. The 'blocking anticyclone', which has been a feature of spring weather in north-west Europe since the mid 1960s, displacing Scandinavian migrant Ospreys to Scotland has been responsible, according to Williamson (1975). He has shown a strong correlation between the rise in spring migrants since 1969–70 and the recent growth in the number of breeding pairs (Sharrock, 1976).

A similar increase in spring migrants (though on a much smaller scale) has been noted in Ireland. Ospreys are now regularly seen in some of the lakes of the west of Ireland. Normally a single bird appears but it often lingers for weeks: unfortunately it is most often sent on its way as a result of being shot at or hassled in some other way. In the past decade two Ospreys have been seen in suitable breeding habitat in more than one location. Given the right (undisturbed) circumstances, it is reasonable to expect them to return to breed here in the not too distant future.

The success story that has been Loch Garten for the past forty years, attracting more than a million tourists, is also possible here, provided the western lakes can sustain sufficient fish stocks to support the hungry birds and that an enlightened attitude can be invested with the people whose livelihood depends on the stocks. It is important, however, to keep a sense of realism about the prospects of repopulation. There is no doubt that our inland waters are, in terms of their productivity, only a shadow of their former selves: over-fishing *in situ*, fish-farming, monofilament nets, arterial drainage, forestry run-off and pollution have all contributed.

The following two historical cameos illustrate vividly the change which has occurred in recent centuries. In 1632, at Coleraine on the Lower Bann, some 62 tons of salmon were netted in a single day (Bardon). If we assume an average of 12 lb. per head, we are talking about more than 10,000 fish! Where in Europe (let alone Ireland) would we find such fecundity today? The records of the Board of Commissioners, carrying out the drainage on the Clare river and its tributaries in the mid 19th-century under a famine relief scheme record a strike among the hundreds of workers employed. A strike in the straitened circumstances of the time may seem remarkable enough, but the cause – that the workers were being fed salmon hauled out of the Clare river, three times daily – paints almost a farcical picture. It leaves us in no doubt, however, as to the abundance and availability of salmon before major drainage was completed in the west of Ireland.

It would be unreasonable all the same to suggest that Irish waters are now unsuitable for breeding Ospreys. The Scottish population is spreading. There are clear signs that they may begin to occupy habitats in south Scotland and northern England – habitats that are no more food rich than those in Ireland. With over 150 young Ospreys now fledging each year in Scotland, the likelihood of migrants stopping off in Ireland increases annually.

In 1997 Ospreys were recorded in no less than eight Irish counties, in all four provinces. The sightings were nearly all in spring, presumably therefore Scottish birds returning from west Africa but included a pair which lingered into summer at a suitable breeding locality in Ulster (BWI & NIBA).

GOSHAWK

Accipiter gentilis

The Goshawk superficially resembles a giant Sparrowhawk, Ireland's most familiar raptor.

Like its smaller relative it has broad, rounded wings and a fairly long tail, the ideal shape for hunting in confined places, in or along the edges of woodland. While the Sparrowhawk preys on smallish birds (finches, Blackbirds) the Goshawk takes mainly medium-sized quarry (crows, pigeons) and also mammals such as Rabbits and Hares. Indeed the formidable, buzzard-sized female can be trained (in falconry) to tackle prey considerably bulkier than itself: the name 'Gos-hawk' is derived from its capacity to hunt and kill geese.

The Goshawk is the classical fierce-looking hawk with staring yellow eye and broad white eye-stripe, offset by dark cheek and cap. Its beak and talons are strongly hooked. The upper plumage is greyish-brown: the lower plumage is white heavily barred dark. The underparts of the immature are thickly streaked rather than barred.

Woodland, particularly extensive undisturbed tracts of forest is the favoured habitat. Throughout Europe this bird is found in both hardwood and softwood forest, with an apparent preference for conifers, particularly in the north. Formerly Goshawks were retiring birds, nesting mainly in the oldest, most remote segments of an extensive woodland. Territories of up to 12,500 acres (5000 ha) containing both forest and open country have been described (Bijleveld, 1974). In recent years, however, Goshawks have shown a remarkable capacity for adaptability. In parts of western Europe they may nest in relatively small copses in agricultural land and have even been found breeding in urban situations where food is plentiful. The large nest is almost always in a high tree and three or four eggs is the usual clutch.

Despite a prolonged history of persecution, the Goshawk survives in suitable habitat throughout Europe. It has even expanded to occupy formerly vacated areas and has recently returned to Britain as a breeding bird, probably as a result of the expansion of forestry. It now breeds regularly, though in small numbers in England, Scotland and Wales. The latest figures (*c*.500 pairs, IRSG, 1999) are most encouraging.

While adults are normally resident and sedentary throughout the year, young birds are partial migrants, mainly to the south.

There has been much confusion about the former status of this bird in Ireland, many believing that the Peregrine was the bird alluded to in the old references. The Peregrine has indeed been mistakenly called by a variety of names including 'Hobby' as late as the early 19th-century (Templeton). The use of falconry terms – tiercel, tarsel (and variations) and gentil, gentille and others – denoting male and female falcon and hawk have, at times, been applied inconsistently, compounding the confusion. References to Goshawks can usually be established as accurate, however, when there is a suggestion of the nest-site being in woodland: Peregrines are usually cliff-nesters.

We know that there have been Goshawks in Ireland for a very long time. The fossil evidence is truly ancient and comes from widely separated localities – Mount Sandel, near Coleraine (*c*.7000 BC), Dalkey Island, Dublin, and Newgrange (*c*.3000 BC). Though hawking has been practised since pre-historic times elsewhere in the world (such as the Orient), these early Irish remains are most likely of pets rather than working birds.

The Irish name is *Seabhac* (pronounced 'shouk') but this is more a generic than a specific name, requiring the adjective to establish identity. O'Sullivan Beare helps us here. Under the heading *Seabhac* he wrote: 'Cuius in Ibernia duo sunt genera: maritimus in rupibus, sylvester in sylvis nidificans et commorans' ('Of which in Ireland there are two kinds: the sea hawk on the cliffs, the wood-hawk which nests and remains in the woods'). As he goes on to name and describe all other birds of prey including the Sparrowhawk, which he defines by the Irish name *speoroig* and the Latin *Nisus*, there can be no doubt that he was referring to the Peregrine and the Goshawk, respectively.

Another, more enigmatic, reference comes from a 12th-century Irish poem. In the extended list of the names of birds is the reference, '*Da sheabhac Feadha Comnach*' ('Two hawks from the wood of Caenach'). As *Spireoicc* (the old Irish word for Sparrowhawk) is mentioned later in the same text, '*Seabhac*' would appear to be Goshawk. The word Seabhac is still in widespread use in Irish placenames, but usually in the corrupted phonetic form shouk, shoke or similar. Many of those that survive to the present day relate to the 'hawk's rock' like Carrick na shoke in Co. Cavan, probably a reference to a Peregrine's rather than a Goshawk's haunt. But some also refer to once-heavily wooded places and may allude to the former territories of Goshawks.

As stated already, the name Goshawk derives from the Old English *Goshafoc*, literally goose-hawk (Lockwood), indicating that the goose was the ultimate prey of this powerful bird. 'It takes not only Partridge and Pheasant but also greater fowl as geese and Cranes' (Rutty). In medieval England Goshawks were indeed trained to hunt Cranes and are depicted thus on an extant Anglo-Saxon picture. At the drawing up of the 11th-century Doomsday Book, careful notes recording the numbers and locations of the eyries of birds of the gauntlet like Goshawks were included. The value of each eyrie, £10 (a very high sum then), showed the importance of the birds in aristocratic life at the time.

At Abbeyknockmoy in Co. Galway a once beautiful wall fresco (now unfortunately gone) depicted 13th or 14th-century Irish kings with hawks or falcons. While the photographic images are too indistinct to define species, their size and lack of striking plumage features (peregrines would show moustache marks) suggests

Goshawks. There is also a passing mention of 'hawks' being flown against Partridges by the Norman sheriff of Louth, Richard Gernon, in 1311 (Calendar of Judiciary Rolls: Edward II). This confirms the practice of falconry in Ireland in medieval times and it is likely that the Normans introduced the sport here, for there is no mention of it in Old Irish. The sport was still a pursuit of the aristocracy until the later 18th century at least; there is a reference in the writings of Coquebert de Montebret to a meeting with Clanricarde's falconers in 1791.

Ranulph Higden writing in the 14th century about Irish birds of prey mentions the 'gentille gossehawke' using the falconry term for the female bird. In Norman law there was a well defined ranking system which restricted certain personages to certain birds of prey. The female Goshawk came high on the list and was regarded as a much more valuable hunting bird than the smaller male.

Due to the value of these birds, especially to the landowner on whose land they nested, Goshawks could be paid to the feudal overlord as rent. There are many references to this practice throughout the Middle Ages. The Finnetors of Limerick were paying dues in this manner as early as the 13th-century.

In the Red Book of Ormond, a 14th-century document from Kilkenny Castle, there is mention of three birds of prey which were used in the payment of rent. The names are in Latin and in two of the cases difficult to interpret. *Nisus* is listed by White (ed.) as 'Fish-hawk', yet this is still the Latin name of the Sparrowhawk. He then lists Goshawk as *Ostorus*, though this word looks closer to the original Latin name for the Osprey (*Ossifraga*) used by O'Sullivan Beare. The third name Esperuarius or Speruarius (listed by White as Sparrowhawk) may be the Goshawk, for it certainly was a rent bird at that time, but I cannot connect the two given the lack of background information in the document and the difference between the normal Latin name *Accipiter* for the Goshawk.

Black-marketeering in the trade of Goshawks from Ireland, mainly to English buyers was rampant. As early as 1218 a Reginald Talbot was seized in Dalkey for illegally delivering an Irish Goshawk to a buyer and was heavily fined. In Richard II's reign a proclamation at Drogheda against exporting birds of prey (1386)

38 *Irish king and hawk (Goshawk?)*
Abbeyknockmoy, Co. Galway

was necessary to reduce the activity and searches were carried out at the ports in Henry IV's reign (*c.*1400) to locate illegally exported raptors. Further legislation was necessary in 1480 in the reign of Edward IV.

The punitive legislation was, of course, in response to the strict protocol regarding falconry: Goshawks were exclusively the reserve of the upper nobility and illegal trade in them was tantamount to monarchistic insult. Conversely, permitting them to be legally taken or sending them as presents inevitably ensured royal favour. As early as Edward III's reign his falconer, William of Troyes, was sent to Ireland to acquire for the king six Goshawks and six 'Tarsals'. By the early 16th century birds of prey were one of the most desirable presents to be sent from Ireland. During Henry VII's time the archbishop of Dublin sent a number as a present to the 'King's Highness'. Two more were sent by the dowager countess of Ormond. The earl of Ossory sent one to Henry's secretary and another to the lord chancellor.

Despite, though perhaps paradoxically as a result of the organised taking of young birds (nests were protected for the purpose of supplying chicks), Goshawks appear to have remained widespread resident birds into Tudor times, in Ireland. In Derricke's falconry poem (1581) the Goshawk is mentioned along with a list of other birds of prey available then in Ireland (see the

Note below). An inventory of Goshawk eyrie sites in the woodlands of north Kerry and Limerick survives from the later 16th century; Glananurlare wood had 'Ayeria accipetrum vocat goshawkes' ('nests of hawks called goshawks'), and there are three others mentioned besides (Westropp). At about the same time a commission was set up to establish the boundaries of the earl of Ormond's estates in Co. Laois. In the course of this work Goshawk nests in the woods of Lowhill and Ballyouskill were recorded by tenants who took the birds as payments to the earl. In the Calendar of Carew manuscripts of the 16th century, reference is made to 'the O'Reyley's [O'Reilly's] of Co. Cavan who were bound to pay her majesty [Elizabeth I] yearly one sound goshawk'. Unspecified hawks were requested at the same time from estates in Cork and from the Macdonnells of the 'glynns' of Co. Antrim.

In discussing the natural history of Co. Leitrim in the 17th century Thomas Molyneux stated that 'the woods are full of large and excellent timber: and well stored with excellent Goshawkes'.

There are in the archives references to the best Goshawks in Ireland coming from the north of the country. One of these, an apparently reliable falconry document, specified that the best Goshawks were to be found 'Especially the county of Tyrone' (Blome, 1686). Ware, in his mid 17th-century *Whole works* recorded that 'Among the feathered kind there breed in Ireland hawks which from their preying on wild geese are called in English gos-hawkes of which those that breed in the North of Ireland are reckoned the best in the world'! Since Glenconkeyne and Killetra on the western side of Lough Neagh probably constituted the largest wooded area in Ireland at that time, it could well be that the Tudor writers were aware of eyries there.

The number of counties in which it was known to nest (Cork, Kerry, Limerick, Laois, Cavan, Letrim, Tyrone; possibly also Meath, Dublin, Derry and Antrim) indicates that, at or around the beginning of the Early Modern period, the Goshawk was a widespread resident here. As would be expected, the above counties were known to be particularly well wooded at that time (McCracken).

The Goshawk was also a widespread resident in the English forests. No less than 19 eyries were identified in the county of

Cheshire in the Doomsday book and Yapp concluded, 'If Cheshire is considered an average county there must have been at the time of writing (11th century) about 1,000 Goshawk eyries in England.' In Scotland, too, Boethius recorded that the Goshawk was one of the resident birds of prey in 1526; some nests still remained in the old fir woods of the valley of the Spey in 1784. The last Goshawks bred in Morayshire and Aberdeenshire in the mid 19th century, after which time they had probably entirely disappeared from Britain.

Apart from Rutty's reference to it, the 18th-century writers are silent about the dwindling status of this bird in Ireland. Templeton (who also illustrated it, leaving us in no doubt as to identity) believed that it might have bred in Co. Derry/Londonderry as late as the beginning of the 19th century. It would appear to have died out completely at about this time, for there are no further references to it other than as a rare visitor.

The gradual disappearance of the Goshawk seems to have corresponded to the falling off in popularity of the sport of hawking from the mid 18th century on. It is hard to believe it now but the sport was so popular in 1641 that it had to be prohibited within seven miles of Dublin (an order of the Irish House of Lords regarding grievances). The Curragh of Kildare was referred to as a preserve in the 17th century, and doubtless the Goshawk was flown at the Hares and other game which were conserved there for the entertainment of the gentry.

Several factors had a bearing on the disappearance of this bird of prey. The fall-off in the popularity of falconry and hawking would have rendered Goshawks (and other raptors) vulnerable: since they were no longer used as rent payments, from the 17th century on, there was no reason to conserve their eyries. This, coupled with the zealous activities of the new demesnes' professional gamekeepers whose brief it was to eradicate any creature that threatened game stocks, must surely have numbered the Goshawk's days. There can be no denying, however, that the natural history of the Goshawk in Ireland was inextricably bound up with the natural history of our native forests. While the habitat remained – at least in viable enough fragments – the Goshawk also remained. Palynology and Dendrochronology have demonstrated

39 *Goshawk by John Templeton, c.1800*

that at certain periods over the past three thousand or so years there was extensive deforestation in Ireland (as elsewhere). While the prehistoric effects were limited by the thin distribution of population as well as technological constraints, the Early Medieval (400– c.1000 AD) effects appear to have been both marked and widespread. It has nevertheless been stated that the extent of primary forest and regenerated secondary woodland was such that the continued survival of our larger fauna (including our avifauna) was not directly threatened. The final (17th-century) clearances, on the other hand, were so sweeping and comprehensive that viable habitats for birds such as Capercaillie and Goshawk and animals such as the Wolf were more or less eliminated. The new industries (ironworks, glassworks) and older greatly developed ones (charcoal production for tanning, barrel manufacture, house- and ship-building) put paid to the remaining forest tracts by the end of the 18th century. In Scotland the story was somewhat different: extensive tracts of mature pine forest remained after the Goshawk had gone; it was probably eradicated therefore by gun and gin trap and, as time has shown, it has returned since the habitat has remained.

40 *Head of Goshawk*

In Ireland Goshawks still turn up but generally as rare visitors. It is debatable whether our present woodlands could sustain them as they once did, for they contain much less game than they did even when the Goshawks' tenancy was coming to an end. They have nested, however, in mature Sitka spruce plantations in Britain and, as we are annually increasing the area of our countryside under this crop, there remains the exciting possibility of the Goshawk's return here in the not too distant future. (A pair of Goshawks bred in Northern Ireland in 1994; there are also unconfirmed reports of the Goshawk attempting to breed in the Republic: IRSG, 1999.)

NOTE

Herewith part of a Tudor poem describing Irish birds of prey used in falconry, from J. Derricke, The image of Ireland *(1581):*

The names of the hawkes that are bred in Ireland with their estimations orderly which are in number seven [?] … Whose names if pacience will abide, in order shall proceede

The Goshawke first of the crewe, deserves to have the name:	[Goshawk, female]
The Faucon next for high attempts, in glorie and in fame.	[Peregrine, female]
The Tarsell then ensueth on, good reason tis that he: for flying haukes in Ireland next, the Faucon plaste should bee.	[Goshawk, male]
The Trasell is gentels course in nexte, the fourth peer of the lande: Combined to the Faucon, with a lovers friendly bande.	[Peregrine, male]
The pretie Marlion is the fifth,	[Merlin, female]
to her the Sparhauke nexte,	[Sparrowhawk, female]
and then the Jacke and Musket laste, by who the birds are nexte.	[Merlin and Sparrowhawk, male]
These are the haukes which chefly breed, in fertile Irish grounde: whose match for flight and speedie wing, elsewhere be hardly founde …	

CRANE

Grus grus

Although it superficially resembles an outsize heron, the Crane is only a distant relative: it is more closely related, in fact, to the Moorhen or the Corncrake. When fully upright, the Crane stands as high as a man's shoulder and its extended length, from bill tip to toe tip, is six feet. It is by any standards a 'grand' bird.

Like the Grey Heron it is generally greyish in colour but with distinctive black and white head stripes and a small unfeathered red crown-patch. The Crane (both sexes) sports a 'bustle' of blackish plume-like feathers (actually elongated scapular feathers) at the rear end which give it a singular profile. Even in the crudest, early depictions this feature identifies the Crane. In flight it is equally distinctive: the long neck is extended, not tucked-back in the fashion of the heron. Also, unlike herons Cranes are gregarious. Flocking occurs at migration times when the music of their wild, far-carrying, trumpeting calls is heard.

Unlike the Grey Heron which usually nests in colonies in trees, the Crane favours extensive and undisturbed wetlands, especially bogs, in which to nest. It relies heavily on remoteness or inaccessibility in its choice of site. The nest is invariably on the ground, in a dry spot in a mire. Two eggs are normally laid.

The present range of the Crane extends from Russia across northern Europe as far as Denmark. The wintering grounds are mainly in Spain, North Africa and Asia Minor, and their traditional migration routes across the Mediterranean have been known and identified since prehistoric times.

Though a wide range of invertebrates and even small animals are eaten, the main food is shoots, fruits and seeds of a range of bog plants. This fact has undoubtedly contributed to the culinary desirability of the Crane through the ages – the flesh of grain-eating birds tending to be most palatable to humans.

The occasional Crane which turns up in Ireland nowadays is usually a straggler from the migratory flocks.

The Crane is one of those birds which has been noticed throughout much of man's recorded history. Its large size, stately poise, gregarious flocking behaviour and seasonal arrival and dispersal have caught the imagination of observers and recorders down the ages. Over two thousand years ago the ancient Greeks noticed and commented on its long-distance migrations. Herodotus (*c.*450 BC) knew that Cranes came from the land to the north of the Black Sea and that they wintered to the south of the Mediterranean. Aristotle (*c.*350 BC) thought that they wintered further to the south, in central Africa.

Documentary evidence survives regarding its inhabitation of England and Wales throughout the Middle Ages (see Notes). It was also known from Scotland in this period. Much of the reference material from Ireland is ancient, though there are a few allusions up to the end of the Middle Ages and even into the Early Modern period (*c.*1600) when it seems to have vanished.

Crane bones were found in the Neolithic/Bronze Age strata during excavations at Lough Gur in Co. Limerick (Ó Ríordáin, 1954). Ballycotton, Co. Cork, is also recorded as a Bronze Age midden site for Crane remains (Cabot, 1999). Others, undated, but perhaps prehistoric have been found in Edenvale cave in Co. Clare (*Trans. RIA*). They obviously constitute midden material of animal or human cave-dwellers. The famous Lagore crannog, royal residence in Co. Meath, was another Crane-bone site. Here they were uncovered amid the remains of other aquatic birds like herons which were presumably also food items. They dated from the 7th to the 10th centuries AD. Other, second millennium, sites include Fishamble Street, Dublin, 10th/11th-cent. (T. O'S, 1991); Dublin Castle, 10th & 13th-cent. (McCarthy, 1995); Wexford, 12th-cent. (McCarthy, 1995); and 17th-cent. from Roscrea castle, Co. Tipperary (McCarthy 1995). This last date is significant in that it suggests that Cranes still occupied the vast bog country of the south Midlands until the beginning of the Early Modern period.

The many bones uncovered from refuse heaps of Roman and Viking habitations indicate the high menu-status of Cranes in early historical Britain.

The high culinary desirability combined with its visual and symbolic appeal have created a corpus of written material from cultures throughout the range of the Crane outweighing that of almost any other bird. There are, for instance, more placenames in Britain associated with the Crane than any other bird. In Ireland too it is remembered in placenames but also in poetry, ecclesiastical literature, folklore, Irish language and visual art.

The word in Irish for the Crane is *Corr*, usually with an adjective appended. Because this appendage is not always consistent, confusion may exist with other Crane-like birds (particularly the Grey Heron), which are also called by the root word (*corr* in Irish).

Humphrey O'Sullivan (1828) discusses this problem in his diaries. He states that the heron has three Irish names but the Crane is simply *Corr monadh*. This name is also mentioned in the medieval Irish translation of Jeremiah 8: 7. O'Sullivan Beare mentions *Coirri moini* (Old Irish equivalent of *Corr mónadh*) in the *Zoilomastix* and ties it down with the Latin name, *Grus*. K'Eogh (1739) helps us further in writing the phonetic interpretation of the name as 'Corrvona', the first letter of the suffix having been aspirated from m to mh ('v' sounding) in the genitive.

In other names the suffix hints at the habitat of the bird in question: *Corr riasc*, heron of the marshes, *Bunnán léana*, Bittern of the reedswamp. In the Crane's case the suffix is *mona(dh)*, the genitive of *móin*, lit., peat (bog) and suggests that the Crane was

an inhabitant of Ireland's bogs, presumably the vast open raised bogs of the midlands.

In the earliest references, however, the name is simply *Corr* and it is left to the reader to determine whether the bird intended is a heron, a Crane or some other similar bird. Armstrong (1946) highlights the problem of identity in focusing on a 14th-century manuscript. A 'Corr' is described as protecting its two nestlings in an open-ground situation from a predatory fox. He makes the point that the bird in question is evidently a Crane due to the circumstance (herons nest in trees) and the fact that Cranes normally have two chicks. The Brehon laws refer to 'Corrs' in a number of places both as wild birds and as pets. Some of the references are enigmatic, even mysterious, assuming a certain knowledge from the reader (often the case in law texts) and point to a cult or ritual relationship between the early medieval Irish and the Crane. The term *Corrguineacht* is referred to as 'Crane sorcery' by Fergus Kelly (1997). He also mentions that the fine for killing a pet Crane (or heron) was disproportionately high, suggesting the specialness of the bird/keeper relationship. The revulsion to Crane flesh among the native (Gaelic) Irish, mentioned by Giraldus, points also to a certain inexplicable distinction in a bird so good to eat. Nearly all the references to Cranes in the Latin lives of the saints are as pets, and it seems that keeping them so (as in early medieval England) was a common practice. In St Adamnan's 7th-century book about the life of St Colmcille, a Crane (*Grus* in the Latin text) is described as seeking him out on the island of Iona. St Abban translates the Irish lake name Loch na Corr as *Stagnum gruum*, lake of the Cranes. One of the most explicit references to wild Cranes, presumably fattening up for autumn migration comes from the Life of St Ailbe who refers to a great flock of Cranes destroying crops (grazing and cereals). Taken in conjunction with Giraldus' mention of their abundance in Ireland in the Middle Ages (flocks of a hundred and more), this scenario is evocative of the present-day conflict between cereal growers and migratory Cranes (up to 7000) which occupy fields in Extremadura in central Spain. In view of the number of Crane placenames in the bog and fen country of the East of England, it

Ardea Grus Crane

Weight 10 lb length 6 feet

This bird was seen in this country during the remarkable hard [...] of 1708 and they [...]

41 *Crane by John Templeton,* c.*1800*

would seem reasonable to look for a similar connection in the Midlands of Ireland. A great many places have indeed been named 'Corr', but this could easily be due to other sources. *Corr* is used, for instance, in the sense of 'odd' to describe a singular topographical feature like an unusually shaped hill or the like. There is a concentration of Corr placenames in the north Midlands, especially in Monaghan and Cavan. Despite the proliferation of wetlands in this region it is impossible to say now that these localities derived their placenames from the former haunts of Cranes. Joyce (1902) thought that, when the forms Cor, Gor, or Gore were associated with wetlands, they applied to 'birds of the Crane kind'. He mentioned a number of examples like Monagor (the bog of the Crane) in Monaghan which, in keeping with the discussion earlier, could well have held Cranes in the past. The 'Crane Islands' of Lough Ramor and Lough Sheelin (Co. Cavan) on the other hand surely refer to herons, the name 'Crane' being a well-known alternative for heron in south Ulster. Indeed, 'Cran' or 'Hern-cran' are widely used folk-names for the Grey Heron throughout Ulster (Braidwood et al.). There may thus have been a transference of the name to the heron on the extinction of the Crane: this kind of transference is recognisable in other pairings.

The Crane has featured as a motif in the visual arts, in artefacts recovered from excavation and in ancient illustrative material (see figs 2–4 on page 17). A garment pin with a Crane-head motif and a crude bone etching, most closely resembling a Crane were found at Lagore: they date from a thousand years ago. Stylised representations probably of this bird, of about this vintage, are to be found in the ecclesiastical text illuminations, complementing the numerous early poetic references. A Viking coin from Wood Quay, Dublin, dating from the 10th/11th-century has an unmistakable stylised Crane on it. A floor-tile recovered from excavations at Mellifont abbey, Co. Louth (13th-century?) is also decorated with a stylised Crane. The finest and clearest artistic representation is a drawing in the original manuscript of *Topographia Hiberniae* (*c.*1200 AD). The bird is shown complete with bustle, striped head and quaintly exaggerated stride. There is no doubt both from the illustration and the text that Giraldus was accurate in his reference.

Ranulph Higden in his *Polychronican* (14th-century) also mentions Cranes in Ireland. Richard Stanihurst writing in *Holinshed's Chronicle* in the late 16th century specifically mentions Cranes breeding in the north of Ireland – 'In the further part of Ulster, certain hils neere to Saint Bean his church [?], where cranes yearlie breed' – and a Tudor contemporary, Payne, also mentions them. The 16th-century Tudor poet Spenser alludes to Cranes in his *Faerie Queene*. Writing about these references, Armstrong says that 'As is well known, many of the details in *The Faerie Queene* show the influence of the poet's sojourn in Ireland from 1580–9 and again in 1597. Much of the poem was written in Ireland and he is more likely to have seen Cranes there than in England.'

O'Sullivan Beare not only authenticates his familiarity with the Crane in his description but he also supplies us with its Latin, Greek, Spanish and Irish names by way of verification. He describes the migratory preparations as follows: 'when they are to journey they gather in a very dense flock: they agree as to the time of departure; they elect a leader whom they will follow, which one neck erect foresees and announces beforehand; those gathered at the end of the flock cry out and keep the flock together with their call …'

The 18th-century naturalists, however, painted a very different picture. Smith stated that 'In the great frost of 1739 some few Cranes were seen in this County [Waterford] but not since or before in person's memory' and 'they do not breed with us'. Rutty writing about the birds of Co. Dublin (1772) mentions only the Grey Heron, as does Harris, in reference to Co. Down. The impression is that the Crane was more or less unknown from Cork, Kerry, Waterford, Dublin and Down in the 18th century. It is not possible, however to extrapolate the absence to the rest of the country, as those counties were mostly unsuitable for Cranes anyway. But if they were still breeding then it is unlikely that these authors would not have heard about it. Thompson in the mid 19th century knew of the Crane only as a rare visitor (but alludes to its past history) and his contemporaries are largely silent on the subject.

It is difficult to explain the disappearance of this once-abundant Irish bird other than in the context of over-hunting for food, not so much by the Gaelic Irish (though archaeozoological evidence shows that they ate Cranes) but particularly by the Normans and other colonists who brought culinary sophistication with them. Certainly a deal of data from Anglo-Saxon and Norman Britain indicates that Cranes were avidly sought – with falcon and crossbow – for this purpose. The primary habitat in England was the vast undrained fen district of East Anglia. Here it is recorded that Cranes were hunted with Goshawks and other birds of the gauntlet until the first firearms in the 15th century ensured their elimination. Willughby (1678) thought that the introduction of the shoulder gun as distinct from the older methods of arrow and hawk was the significant factor in their disappearance from England. He mentioned then that Cranes had not bred in England for over two centuries. Others have suggested that Willughby's information about breeding was incorrect. Turner (1544) stated that the Crane still nested in Britain then and there is a 16th-century reference to its breeding in Wales. There is evidence that it was also nesting in Scotland at that time. The 'Scotch Act' of 1551 fixed the price of a Crane at five shillings (a fair sum then), indicating its continuing desirability for the pot despite increasing rarity. Even as early as 1275, at the royal table of permitted poulterers it was declared that a Crane should cost three shillings as against sixpence for a heron or a Bittern. At a banquet held in Dublin in the 12th century for Henry II Cranes were on the menu; Cranes and other large game birds were among the favourite dishes of the Normans, and history records that this event may also have been a changing point in the native taste: 'Throughout the great hall in obedience to the King's [Henry's] wishes, they began to eat the flesh of the Crane (*carne gruina*) which they had hitherto loathed' (Giraldus Cambrensis, *c*.1185). At a great feast during the reign of Edward IV it was reported that 204 Cranes were served. The Crane's diet of grain, fenny seeds and bents rendered their flesh highly palatable, regarded by Moffett in the 16th century as 'esteemed good eating, far superior to the Heron'.

The late medieval drainage operations on the wetlands was put forward as a primary factor in the Crane's disappearance from England. It certainly could not be put forward as a reason for the

42 *Pair of Cranes and young*

loss of the Crane in Ireland, for drainage did not impact on our wetlands until much later (mid 19th century), by which time the Crane was long gone. In fact, if we accept that the raised bog was the main habitat of the bird in Ireland, it was in theory possible for it to survive here until the beginning of the 20th-century exploitation, for non-mechanical bog-cutting (as shown in the early 19th-century Bog Commission maps) was entirely peripheral to these vast habitats. So, habitat destruction was certainly not the deciding factor. Hawking was widely practised in Norman Ireland, and the Crane was the ultimate quarry for the falconer, here as elsewhere in medieval Europe. It is therefore reasonable to deduce that falconry took its toll of the Crane population, particularly in the East and in the Midlands. However, even taking into account the desirability of this bird for hunting it should surely have survived as long as, for instance, the Capercaillie, which was also esteemed for the pot. It could be argued that the Crane, being a large, vociferous, open-country bird, is especially noticeable and therefore vulnerable, but so are geese and swans and they have managed to survive despite centuries of hunting and disturbance.

It is likely therefore that other factors were involved in the disappearance of Cranes from Ireland and climatic change (the little Ice Age?) seems as likely a candidate as any.

Just how common they actually were in historic times is worth considering. Though described as 'abundant' in early accounts, Cranes migrate *en masse*, and a thinly distributed population could give the impression of abundance at migration times. After the decline and eventual disappearance of the (local) British population (around the turn of the 16th/17th century) the Irish population would have remained as an isolated fraction on the extreme western edge of its European range. It has been shown that in these circumstances migratory island populations are unstable and more prone to vacate their marginal territory than sedentary species. Cranes have, as an extension of this phenomenon, vacated all their western European breeding grounds over the past few centuries, though a small fraction (mainly from Scandinavia) a relic of what must at one time have been a large western population, still migrates annually to, and through, Spain and Gibraltar to Africa.

The natural history of the Crane in Ireland is an unfortunate one for us. In its disappearance we have lost an integral part of what was once an essentially Irish wilderness phenomenon. The American nature writer, Aldo Leopold, conveys a haunting sense of the 'crane-bog' in a *Sand County Almanac* (1949):

> To the residual lagoons came the cranes [Sandhill Cranes], bugling the defeat of the retreating winter, summoning the on-creeping host of living things to their collective task of marsh-building. Floating bogs of sphagnum moss clogged the lower waters, filled them. Sedge … advanced over the bog, anchoring it by their root fabric, sucking out its water, making peat. The lagoons disappeared but not the cranes. To the moss-meadows that replaced the ancient waterways they returned each spring to dance and bugle and rear their gangling, sorrel-coloured young.

Thankfully we have been wise enough to retain elements of the original habitat, extensive enough to convey some idea of their

43 *Cranes taking to flight*

original magic. But the Cranes are long gone. Given the pressures which have been brought to bear on all large creatures with extensive habitat requirements, it is difficult to see now how its disappearance could have been avoided. The situation is, however, not without hope. Cranes continue to turn up in Ireland as occasional and rare visitors, often in winter. On occasions small flocks have appeared, and individual birds have lingered into the breeding season (Cranes have summered regularly in eastern England in recent years, and one pair has bred and successfully raised young from 1981 to at least 1988: *Red Data Birds in Britain*, 1990). With Bord na Móna gradually winding down operations over much of the cut-away bogs in the midlands, it is likely that there will be a strong lobby from conservationists to have these otherwise unusable 'wastelands' set aside for wildlife. With such suitable wetland habitat being made available on an increasing basis, we must wait in hope that the Crane may once return to grace our boglands.

NOTES

Cranes as pets: 'Fosbroke tells us that "it was the custom in the Middle Ages to keep tame cranes which stood before the table at dinner, and even kneeled and bowed the head when a bishop gave the benediction". Whether they could have been made as tractable as this we will not pretend to say: but there seems to be no doubt that they were kept in courtyards and about the houses formerly, just as peafowl are kept at the present day': *Recreations of a Naturalist* by J.E. Harting (1906).

Crane trapping in Pakistan: 'Using lead-weighted cords and caged decoy cranes, hunters are thought to live-catch at least 1500 to 2000 a year. It is believed that hunting pressures may ultimately threaten the more numerous common crane (*Grus grus*) … Most of the captured birds are used as decoy cranes in existing camps or to stock new camps for the growing number of crane hunters. About 10–15 per cent are sold to people who keep cranes as pets and "watchdogs" (Cranes are as exceedingly territorial in captivity as they are in the wild)': 'Crackdown on Crane-killers', *BBC Wildlife Magazine* (May 1985).

The former status of the Crane Grus in Britain: (Ibis 140:482–500): 'There appear to be nearly 300 place-names which include some reference to Cranes very widely distributed across Britain; at least half of the sample has the name associated with other place-name elements relating to water (e.g. fen, mere, lake). No other wild bird appears in so many place-names. Crane bones are also quite common in archaeological sites, though they are absent from most cave sites; they are reported from at least 78 excavations. The evidence of bestiaries, illustrated manuscripts and other documentary sources makes it clear that the Crane was a well-known bird, clearly distinguished from the Grey Heron *Ardea cinerea*. All three lines of evidence confirm that the Crane was a breeding bird in Britain, not just a winter visitor': S. Boisseau & D.W. Yalden, 1998.

CAPERCAILLIE

Tetrao urogallus

The Capercaillie is a massive Eurasian game bird almost as big as its American counterpart, the Turkey. The cock, much larger (about 3 feet, 0.9 m) than the mottled brown, grouse-like hen (2 feet, 0.6 m), is unmistakable. Its plumage overall is sooty slate with a blue-green sheen on the breast. It also sports white shoulder spots and beak and red eye wattles. The sombrely coloured cock is transformed into a demonic spectre by nuptial prospects in the breeding season: it struts on a tree bough or on the ground – neck stretched upwards, throat swollen and tail fanned in a huge, perfect semi-circle – master of its forest domain. Add to this its utter contempt for trespassers, human or otherwise, and its most unbirdlike song (more like a galloping horse) and we are confronted with one of the most spectacular creatures of the forest.

The Capercallie's food comprises the shoots, buds and berries of a variety of trees and woody plants with a marked preference for conifers. Though it is found in mixed deciduous forest in Spain and in parts of temperate Russia, the Capercaillie's prime habitat is coniferous forest, particularly Scots pine with heathy understorey (Bilberry and other woody species). In winter it wanders to scrub and even open ground adjacent to the forest edge. Like others of the grouse family, the cock establishes a harem of females at the breeding ground (a lek). The nest is on the ground: half-a-dozen eggs are laid and incubated by the female. The male takes no part in the rearing of offspring.

Despite its arresting nuptial behaviour and its obvious suitability for the pot, the Capercaillie thrives in forests of the Pyrenees, the Alps and parts of northern Europe, particularly Scandinavia. Also former native of Britain, it was finally eradicated from Scotland in the late 1700s.

Until relatively recently the question of the Capercaillie ever having been an Irish bird was the subject of much debate. In fact, until its bones were found at an Irish archaeological site in the 1980s, the consensus (among those who wrote about it) was that the bird never lived here in the wild: this conviction was born as much from the difficulty in establishing a recognisable name by which it had been formerly called, as the perceived unsuitability of the forest habitat in this country.

The discovery of Capercaillie bones at the Mesolithic settlement at Mount Sandel (near Coleraine) in 1982 resolved the question: the Capercaillie, in prehistoric times at least, had been an Irish native bird. The discovery location lies in what must have been then the thickly afforested Bann river valley, not far from the sea – a likely landing/temporary encampment for hunter/gatherers. Other bird and animal bones uncovered at the excavation (Woodpigeon, Woodcock, Goshawk, Wild Boar etc.) showed that these people had exploited a variety of the game of their immediate surroundings as food (Wijngaaden-Bakker, 1989). Evidence from late Boreal times does not, of course, imply that the Capercaillie survived into Early Modern times, which is now claimed. It would be quite reasonable to believe that, along with the Brown bear and perhaps the Wild Boar it was hunted out before historic times. However, recent bone evidence from a cluster of additional sites: Viking Dublin's Fishamble Street; Norman strata in Dublin castle (10th & 13th-century); Waterford

(12/13th-century) and Wexford (12th-century), clearly point to the Capercaillie's survival into later medieval times (*per* McCarthy & O'Sullivan, University College, Cork). There remains uncertainty about the so-called 'Turkey-bones' found in Irish caves and referred to by Greenoak (1979). An effort has been made to have them specifically identified and dated (*per* Hall) and those re-examined proved indeed to be Turkey bones as originally stated. They must be therefore, of relatively recent origin for Turkeys were not known in Ireland prior to the last decade or so of the 17th century (they were introduced into Lord Conway's Co. Antrim estate at Portmore at about that time), which would suggest that further re-examination might prove fruitless. How they ended up in caves is far from clear, unless they were brought there as prey by foxes or other predators. The wild turkey is mentioned in the Irish game laws of 1787: 'To take, kill, sell, expose or buy … wild turkeys from 10th January to 1st September (a fine) not above £5 a head'.

Capercaillie bone from late Celtic or Roman times (*c.*2000 years ago) has been found in England. Remains have also been found in Somerset and in Northern England (York and Durham) suggesting a formerly wide range there, possibly into historical times. The pre-decline range, therefore, would appear to have covered the whole of the British Isles, though it must have died out in England (and Wales?) prior to later medieval times when bird names were widely recorded in the literature. The Welsh name was apparently *Ceiliog Coed* ('Wood hen') and was mentioned in a document concerning land tenure payment to the bishop of Durham in 1343 (Gurney, 1921).

The name Capercaillie is obviously Gaelic in origin, the latter part being a corruption of the genitive of the Gaelic word *coill* meaning wood. The literal translation is most likely derived from *capall coille* ('horse of the wood': Lockwood, 1984). One explanation for this enigmatic name is the bird's large size, horse being descriptive of bigness as in Horse Mushroom or Horse Fly. A more convincing (and usually accepted) explanation is the 'song' of the male Capercaillie during display which sounds just like the accelerating 'clip-clop' of a galloping horse. The name must be Scots Gaelic as it does not appear to have originated in Ireland:

there is no word or combination of words resembling *capull coille* in any ancient or modern Irish source (*per* RIA). This suggested derivation is not unanimous, however. Saunders (1927) thought that the source word was 'from the Celtic *gabur*, a goat – with allusion to the elongated and beard-like chin-feathers of the male as well as his amorous behaviour in spring … but some authorities prefer *cabhar* an old man, or *gobur* a horse'.

The established, widely used names for Irish birds are very old. The name of the Woodcock (*Creabhar*), for instance, is mentioned in the early literature of perhaps 1400 years ago and due to the much older oral tradition is probably prehistoric at source. It is mentioned repeatedly throughout the Middle and Modern Irish, perhaps because it ranked high on the list of quarry species for hunting.

When Dinneen and Mc Kenna applied the name 'Woodcock' to an old Irish name for an unknown bird *Coileach Feadha,* they were simply translating literally. However, there can be no doubt that the bird intended was a woodland bird other than the Woodcock, for the earliest reference (a 9th-century piece attributed to Flann Fina: *per* Meyer, 1909) elevates it to a status far superior to that which the lowly Woodcock might attain: 'Who are the proudest you have met? … The men of Muskerry and the cocks of the wood (*Coilig Feadha*)'. What other woodland bird could evoke such comment about its demeanour and behaviour? The name was also mentioned in a 12th-century poem from the Duanaire Finn and subsequently in several medieval manuscripts. *Coileach Feadha* was also applied to the Pheasant by Dinneen and McKenna, but since there is no record of the Pheasant in Ireland before the later 16th century, it could not be that bird either. In view of the practice of name transference in the Irish it is reasonable to postulate that the Irish name was transferred to the newcomer with the gradual disappearance of the native bird: both are woodland game birds. Could the fact that they bear no physical resemblance to one another have been sublimated for the sake of other mutual characteristics?

The issue of the name was given some consideration by P.G. Kennedy SJ (*Studies*, 1944): see the Note below. It has also given

G. D'ARCY - 85

Capricalca famina.

44 *Female Capercaillie by Sir Robert Sibbald, 1684*

rise to considerable ornithological controversy, first from C.D. Deane who, in a memorable article in *Irish Birds* (1: 364–9, 1979) asserted that, according to his researches, the Capercaillie was never an Irish bird. A cornerstone of his case was the absence of the Scots Gaelic name in the Irish literature. Deane's findings were challenged by J.J. Hall, also in *Irish Birds* (2: 38–47, 1981). His research indicated that the Capercaillie was indeed a native Irish bird which had gone by the name 'Cock of the wood'. The controversy eventually resolved itself, to a large extent, with the discovery of Capercaillie bones from the Mesolithic site at Mount Sandel. Others have expressed misgivings about the survival of the

Capercaillie into historical times, but the weight of corroborative evidence is so compelling that it is now accepted that the Capercaillie was native and survived in Irish woodlands till Early Modern times at least. O'Sullivan Beare (*c.*1625) referred to the Irish name but seemed to be unclear about the bird in question. He translated it as *Capella banelus* and *Phasianus* (Latin) but he did not intend the Pheasant as he translated that as *Coileach cruaigh* and described it convincingly. He mentioned that there was a turkey-like bird, 'an inhabitant of the dense woods' of Ireland, which he had not seen; this probably accounted for his inability to apply the name correctly.

K'Eogh (1739) mentioned a variation, *Cuelagh cuili* (a phonetic interpretation of the Irish name) which he gave as 'Cock of the wood'. Furthermore he described its alleged medicinal qualities as if from first-hand knowledge of the bird.

The cumulative literary evidence (from both Irish and English sources) as presented by J.J. Hall demonstrated convincingly that the Cock of the wood was indeed the Capercaillie. One of the counter arguments raised was that the pine wood habitat, the primary habitat of the Capercaillie elsewhere in the world, did not survive in Ireland into historical times. The palynological evidence certainly points to widespread, continuous decline of pine from Bronze Age times through Iron Age and medieval times. However, it must still have been common enough to be familiar and available in early historical times since it was listed as one of the seven 'nobles' in the Brehon laws and the *Ochtach* (pine) was mentioned as the tree used as a ridge timber in the construction of a Gaelic chieftain's house. Some of the evidence for regional disappearance is more convincing than others, but since 'firs' were referred to by Tudor and Jacobean commentators it seems reasonable to consider localised pine survival on uplands and bogland at least into Early Modern times. There is an intriguing reference to pines growing in a forested region at the foot of the Sperrin mountains, to the west of Lough Neagh as late as the early 17th century. 'There is no shortage of pine, especially in Conkeny [Glenconkeyne] forest, a quite tall tree on which the biggest nut of all, the pine cone grows hanging very high up, composed of kernels very small on the inside, but greatly pitted on the outside, each covered by a very hard coat. From that tree flows resin, a liquid that is sticky and useful in many situations and today it grows in Ireland in several places in abundance' (O'Sullivan Beare *c.*1625). Waring (1705) referred to 'Firrs' growing in Connaught and eliminated the possibility of mistaken identity by comparing them with those of Scotland.

However, the absence of pine does not necessarily exclude the Capercaillie. It is known to occupy hardwoods also. In parts of Spain and south-eastern Europe (where there are relict populations) and in Russia the Capercaillie is found in mixed hardwoods, particularly oaks. Hall stated that it exists locally in broad-leaved woodland, with holly supplying the evergreen element. He further stated that species on islands (such as Ireland) 'not infrequently' occupy broader ecological niches than the same species do on mainlands (from Lack, 1969). One can imagine therefore the gradual adaptation of the Capercaillie from conifer forest (in pre-historic times) to mixed hardwoods. It would have been more vulnerable in the accessible tracts of mixed forest and would have vanished rapidly during the periods of Plantation in late Tudor and Jacobean times but may well have survived in wilder upland regions, well into the Early Modern period. There is one tantalising reference from Killultagh estate, at the north-east corner of Lough Neagh, Co. Antrim, which appears to bear this out. 'The pheasants are increasing very slowly but there are plenty of cocks of the wood and poults on the upper ground' (G. Rawdon, 17 Nov. 1637: *Calendar of state papers, 1633–47*). This is 'game talk'. The Pheasants had been recently introduced, but despite their slow rate of increase (due to depredation?) the Capercaillie and Grouse were still common enough to be hunted in the wilder uplands, presumably the eastern fringe of the Antrim plateau.

This, of course, was only one allusion to the Capercaillie as a quarry species; such a large fleshy grouse must have been avidly hunted for the pot since earliest times. The Mesolithic Mount Sandel

45 *Capercaillie? in Galway Corporation Book, 1638*

bones were the discarded remains of meals as were those from the Early Medieval sites elsewhere. In one of the earliest references to the *Coileach feadha* (from a 12th-century verse about Diarmuid and Gráinne) the celebrated lover tells his loved one that the flesh of the Cock of the wood taken with a little mead is to be preferred to ruling over a kingdom (Stokes, 1899). As Kelly and Fox point out in their discussion on the Capercaillie (*Irish Birdwatching*, summer 1995) the flesh of the Capercaillie from pine woods is regarded as being 'something of an acquired taste' due to the resinous flavour imparted by the bird's coniferous diet. The Capercaillie from medieval Ireland, however, being a bird adapted to hardwoods (as discussed), probably tasted fine, as do most game birds which feed therein.

There is a cluster of references from the late 16th to the 18th century, a period of particular culinary indulgence, where the sole reason for mentioning the bird was in regard to its suitability for eating.

Holinshed in his *Irish chronicle* (1586) refers to 'Woodcockes', but judging from the emphasis that he places on their uniqueness it is difficult not to believe that it was Cocks of the wood and not Woodcocks that he meant: 'But your lordship who is thoroughly acquainted with the worthiness of the Island [Ireland], will be some persuaded to leave such quaint and licourous repastours [*sic*], to feed on their costly and delicate Woodcockes, and willingly to accept the loving present of your hearty welwiller'. Hall (*Irish Birds*, 1981) quotes two 17th-century references to the 'fine flesh' of the Irish Cock of the wood. There is an interesting reference, uncovered recently (*per* W. McDowell) in *Ibis* 90 (1948). A short note to the Royal Society of London, dated 26 October 1687, read as follows: 'desired that the cappercail or cock of the wood should be inquired after from Ireland, where Dr Mullen said that they were commonly found in the market'. As Dr Mullen, alias Molines (*Manuscripts from the British Museum*, 1933) was Dublin-based, the market referred to was presumably Dublin market. Could it have been, therefore, that the Capercaillies on sale there were obtained from the Dublin or Wicklow mountains, probably the nearest suitable habitat at that time?

Dr Charles Smith in his book *The state of the county and city of Cork* (1749) also comments briefly on the Capercaillie's suitability for the table: 'The flesh is greatly esteemed.' But he tells us more: 'The bird is not found in England and now rarely in Ireland since our woods have been destroyed.' Willughby (1676), who confirmed its absence from England, leaves us in no doubt about how highly regarded it was: 'The flesh of this bird is of a delicate taste and wholesome nourishment, so that being so stately a bird and withal so rare, it seems to be born only for Princes and Great men's tables.'

J.J. Hall's meticulous article on the Cock of the wood in *Irish Birds* (1981) mentions other references to the bird's culinary attributes: 'remarkable for its fine flesh and follie thereof' (Fuller, 1649); 'a most stately bird never known in England, called the Cock of the Wood, as large as a Turky[*sic*] – cock, with black feathers, scarlet eyes, and flesh more white and delicate than a Turky's; a rarity even in Ireland' (Thornton? *c*.1640: *per* Comber 1778). The primary importance of Hall's article was to establish the fact that the Cock of the wood was in fact the Capercaillie: it was clearly not called by the latter name in Ireland. While it contains fragments of information which can help with the reconstruction of the bird's final phase as a member of our wild avifauna, it is not necessary to reproduce the entire article.

We must review the scatter of historical references to try to envisage the pre-extinction distribution, setting them also in the context of the landscape of the time since the Capercaillie must have been more or less confined to the mature woods and forests. Military maps of the Tudor and early Jacobean period show that none of the four provinces was devoid of suitable habitat though mid Ulster and south Munster were particularly so. If we return to Giraldus for our first (tentative) reference, his 'wild peacocks' abounded in the west of Ireland. This is, to an extent corroborated by O'Flagherty, a 17th-century Connaught writer, who wrote: 'I omit other ordinary fowl and birds as … cocks of the wood … ' Despite, apparently not having first-hand knowledge of this bird in his west Cork/Kerry homeland, O'Sullivan Beare did know of it and was able to describe it. The 'men of Muskerry' reference is unambiguous and clearly from Cork. The Thornton, *c*.1640 reference above was from Castlecomer, Co. Kilkenny, which extends the range towards the Midlands. The reference attributed to Dr Mullen from the Dublin market relating to the later 17th century suggests that the birds were hunted in the Dublin or the Wicklow mountains. The dearth of historical information relating to Ulster is strange, considering that this was probably the most afforested and wild region of the whole country before the Plantation. The reference to Lord Conway's estate in 1637 ('plenty of cocks of the wood on the upper ground') certainly suggests that it was not a rarity. There is also a strange anecdotal reference to a 'peacock's feather' being worn in the cap at the inauguration of 'The O'Neill' at Tullahogue in Co. Tyrone in the 16th century; could this have been a Capercaillie's feather? it would certainly be an appropriate symbol of status and pride.

Slim as this scattering of locations is, it nevertheless suggests a fairly broad distribution prior to the final disappearance.

In consideration of the decline it is best to ignore the fossil evidence (both prehistoric and medieval) since they have no bearing on the circumstances in the Early Modern period. The 18th-century references invariably mention the decline or impending disappearance of the Cock of the wood. In an act of the Irish parliament of 1711 it is stated that 'the species of the cock of the wood, a fowl peculiar to this kingdom is in danger of being lost'. Sir James Ware wrote as early as 1658 (according to Harris, his translator, 1745) that he believed the Cock of the wood was at least scarce in Ireland if not extinct. Smith in 1749 made the following brief comment: 'Now found rarely in Ireland'. Rutty (1772) stated that one was seen in Co. Leitrim about the year 1710, 'but they have entirely disappeared of late'. Dr Patrick Browne, Dublin physician and naturalist, included the Cock of the Wood/ Capercaillie in his *Catalogue of Irish birds, fishes and marine mammals* as late as 1774. Pennant noted that 'a few are still to be found about Thomastown, Co. Tipperary [Co. Kilkenny]', in 1760 but Thompson thought that the Capercaillie had become extinct here by about 1750. In the *Proceedings of the Dublin Natural History Society* (December 1851) there is a reference to 'a very few surviving in Ireland until about 1790': Kennedy, Ruttledge and Scroope thought enough of the reference to include it under Capercaillie in *Birds of Ireland* (1954).

In the O.S. memoirs (Donegal) of 1831–32 under 'extirpated gallinae' is mentioned 'cock of the wood or cappercaillie'. It would be impossible to trace the background to this comment now, but the author would hardly have gone to the trouble of recording this had he not been aware of it being formerly found (perhaps in living memory) in that county. The above references somewhat extend our knowledge of the range in the final period prior to its disappearance.

There are no Early Modern references to the Capercaillie in England or Wales, though, as mentioned earlier, it was found there in early medieval times. However, a Forestry Commission leaflet (HMSO, No. 37) states that it did not die out completely in England until 'around 1660–70' (C.E. Palmar, 1965). In *Ornithologica* by F.J. Willughby (1676) we read under 'Capercaillie': 'This bird is found on High Mountains beyond Seas and we are told in Ireland (where they call it cock of the wood) but nowhere in England.' The translated and enlarged version by J. Ray (1678) contained the following. 'The COCK OF THE WOODS or Mountain, Urogallus feu Tetrao major. Aldrov. This is not found in England, but in Ireland there be of them.'

Hector Boethius (1526) mentions it as a Scottish bird, and John Leslie refers to it as being an inhabitant of Ross and Inver in 1578. The actual decline and disappearance of the Capercaillie in Scotland has been well documented and appears to have occurred at roughly the same time as in Ireland – the later 18th-century (Pennant et al.).

Gradual destruction of the Caledonian forests and resultant ease of hunting access are cited as the main reasons for the Scottish disappearance. There is no doubt that the wilder reaches of the Scottish highlands were subjected to concentrated free-for-all exploitation towards the end of the 18th century and into the 19th.

In Ireland, too, Rutty and Smith thought that habitat destruction rather than hunting was the major cause of decline. Having adapted to a habitat shift from conifer to hardwood, the Capercaillie was incapable of adapting again.

It was for hunting purposes that the Capercaillie was reintroduced into Scotland in 1837. Lord Breadalbane imported from

46 *Capercaillie pair in flight*

Sweden thirteen cocks and nineteen hens for the purpose. The introduction was a success and by the early 1860s there were between two and three thousand Capercaillies in or around the Taymouth estate in Perthshire. Nowadays this massive game bird is once again occupying a considerable portion of the territory it formerly occupied in its heyday. It is now carefully conserved (though also hunted) and the future of its forest habitat is now assured.

There were at least two attempts at reintroduction into Ireland in the 19th century – to Markree (Sligo) and Glengarriff (West Cork). Both were unsuccessful, and the celebrated Capercaillie or Cock of the Woods remains an absent member of our indigenous avifauna to this day.

NOTE

From P.G. Kennedy SJ, *Studies*, June 1944, p. 249:

> The large game bird, the Capercaillie or Cock of the woods, has only an historic interest for us as it has long been extinct as an Irish species. Giraldus Cambrensis writing at the end of the 12th century stated that 'wild peacocks' abounded in the Irish woods. They were still common in the 17th century but in the 18th century they gradually disappeared. Dr Charles Smith in his history 'The state of the County and City of Cork' which was ready for publication in November 1749 writes of the Capercaillie 'called in Ireland the

Cock of the Wood: its bigness is near to a turkey … the bird is not found in England and is now rarely in Ireland since our woods have been destroyed. The flesh is highly esteemed.'

Though *coileach feadha* is not explained as Capercaillie in any dictionary and though Dinneen & McKenna give it as a name for both pheasant (*phasianus*) and woodcock (*Scolopax rusticola*) it would nevertheless seem likely that the Capercaillie is the bird meant in the 12th century poem vii, 9.21 (and also in the 12th century MS, ll. 145b 16 *Cailig fheda* [nom. pl.] *man fid ngluair*). First the common word for woodcock is *creabhar* … it is spelt *cruowr* and is defined as woodcock by the 18th-century medical writer John Keogh in his *Zoologica*); and it could well be that the explanation of *Coileach feadha* as woodcock is the result of literal translation.

Next, the explanation of *Coileach feadha* as pheasant certainly cannot hold for a 12th-century poem:

> There is no mention of the pheasant in Ireland till 1589 [*Studies*, art. cit., p. 250]. These statements are corroborated by the fact that in the 15th century there would seem to have been no recognised equivalent for Latin *phasianus* 'a pheasant'. For in the translation (made perhaps in the first quarter of the 15th century) of the Regimen Sanitatis of Magninus Mediolanensis (ed. O'Ceithearnaigh) *fasianorum* 'of pheasants' has been represented by an obvious borrowing from the Latin … When the capercaillie disappeared it may well be, however, that the name *Coileach feadha* was genuinely applied to the pheasant (Keogh i.e., gives *cuelagh-fa* as the Irish for the pheasant) and Fr McGrath has doubtless rightly translated *an creabhar, an c[h]earc f[h]raoich, an cuileach feadha* as 'the woodcock, the grouse, the pheasant' in his edition of H. O'Sullivan's Diary [13 November 1828]'.

GREAT AUK

Alca impennis

The auks have often been described as penguins of the north. The true penguins are not directly related to the auks, are found only in the southern hemisphere and are, of course, flightless. The Great Auk was not only by far the largest of the family (3 feet; 0.9 m long) but was also the only one like the penguins which could not fly. Though they retained the normal structure and feather arrangement, the small wings were mere flippers, held like redundant limbs at the sides of the standing bird. The stance was particularly upright, making movement on land awkward and rendering the auk vulnerable to predators. In water, however, the Great Auk came into its own, swimming both on and beneath the surface with consummate ease. There are a number of eyewitness accounts of the agility and speed of the Great Auk in underwater pursuit of food.

The food was presumably small and medium-sized marine fish (sandeel-sized to herrings), though it is likely that other surface-living creatures (such as squid) were taken too. The Great Auk's breeding range was restricted to the North Atlantic – from the Canadian Arctic on one side to the fringes of northern Europe on the other. Remarkably, it is now believed that they were at least partially migratory – despite being unable to fly! They were known in winter from as far south as Florida on the western Atlantic and the Mediterranean on the European side.

The actual nest sites were on favoured and traditional rocky islets and isolated rock stacks. Here they were found, as with other auks like Guillemots and Razorbills, in considerable density. However, unlike the other species which could occupy niches high up and out of harm's way, the flightless Great Auk was restricted to the accessible, spray-swept lower ledges. As with other auks, no nest was made, the single very large egg being laid directly on to the rock. Unfortunately virtually nothing is known about the rearing of the young bird, though presumably there were parallels to the Guillemot, which also produces only one young.

The Great Auk is the only one of our hapless itinerary which is extinct throughout the world. The scientific and cultural loss is therefore of global rather than simply national significance.

As the only flightless seabird in the northern hemisphere, the evolutionary ramifications are unique and probably will never be fully understood, since the Great Auk was not studied to any degree before its demise. The best that we can now do is to pick through the surviving evidence in reconstructing the lifestyle of a truly remarkable bird.

The Great Auk or Garefowl as it was widely known was also called the Penguin, even before the name was applied to those other, well known flightless birds from the Antarctic. Penguin was thought to derive from the Latin *pinguinus* meaning 'fat', but this has been disputed. Speculation of this kind is understandable, however, given the bird's connection with man as an important food item. Indeed, the evidence shows that its colonies were a 'fat-store' for coastal dwellers and seafarers for a very long time.

Remains from kitchen middens in numerous locations along the coastline of western Europe have shown that Great Auks were being slaughtered for food six thousand and more years ago; Scharff (1899) was of the opinion that the Great Auk was a typical Arctic species (like its diminutive Arctic relative, the Little Auk) which had spread southwards to inhabit the coasts of temperate western Europe and the Atlantic coast of the north American continent. Here it would have been regarded as marine 'manna' by coastal dwelling hunter/gatherer communities. It certainly formed concentrated populations in a number of prehistoric sites around the coast of Scandinavia and Britain. In Ireland its bones have been uncovered from prehistoric middens on the coasts of Cos. Antrim, Donegal, Clare and Waterford and also in caves in Waterford. In some of these sites the bones were described as 'abundant', reflecting perhaps the density of the colonies. The bones are large and highly recognisable amid other midden refuse. It is possible that other ancient dining places (*fulachta fiadh*, for instance), of which there are hundreds around our coasts, as yet unexcavated, have also Great Auk bones which would reveal additional information about its former range in Ireland.

The Great Auk's former world range has been the subject of considerable speculation. Lucas (1891) stated that it was confined to the North Atlantic ranging on the European side from Iceland to northern Spain and on the American side from Greenland to Virginia. This, however, probably represents a reduced range, for it is known in an ancient context as a widespread species. Its bones have been found in Pleistocene deposits in Italy dating from some 60,000 years ago and Armstrong (1958) remarked that 'remains of Great Auks slaughtered by Neanderthal man occur as far south as Apulia and Gibraltar'. There are several late Ice Age drawings in a cave near Marsailles, which have been interpreted as Great Auks. This southern evidence was probably from wintering birds for, remarkable as it may seem, at least some Great Auks were long-distance migrants though they must have swum all the way! Post-breeding dispersal from the scattered northern colonies seems to have taken the Auk to the shelter of the Mediterranean on one side of the Atlantic and the Panhandle region on the other. This raises the interesting question: Were Irish Great Auks resident, that is, nesting around our coasts, or were they migrants on their way further south for the winter? Breeding colonies were known in historical times on rocky stacks off the north American coast, including Funk Island, the bay of St Lawrence, Cape Breton, the coast of Iceland, possibly the Faroes, the Shetlands, Orkney and St

Kilda. Armstrong considered that the bones from kitchen middens at Oronsay and those from Whitepark Bay, Co. Antrim, had to be from birds caught at breeding sites, indicating that the species nested further south in post-glacial times than in the period of recorded history. The sheer abundance of bones recovered from the Irish sites indicates the pillaging of breeding colonies – not the random trapping of migrants. There were certainly suitable breeding sites near the middens such as the Skerries near Whitepark Bay. Ussher was of the opinion that the Great Auk may have nested on the Keeragh islands off the Wexford coast, the nearest suitable habitat to the Waterford coast, where its bones have been found in quantity. A thorough archaeozoological investigation of this exposed reef might prove most interesting.

The search of the historical literature reveals surprisingly little reliable information about Irish Great Auks. There are tantalising bits and pieces which do no more than whet the investigative appetite, but clear first-hand description is not forthcoming. There does not appear to be an Irish name, for instance. There is a Scots Gaelic name, however; in a description of the Western Isles of Scotland by Martin (1716) there is a reference to the Great Auk which is called 'Bonnivochil' or 'Bishop-fowl'. The former name is obviously a phonetic interpretation of the Gaelic name. On St Kilda it was known to the islanders by the Gaelic name *Bunnabhuachaille* (Heathcote, 1900) which is more or less the same. This name is a derivation of the Gaelic *buachaill* (boy) with presumably, a prefix of endearment. The meaning overall might approximate to 'stout little lad' or 'cocky little fellow'. One could imagine this name also being used in Ireland but, as the Capercaillie case has shown, there need not be consistency between the Irish and the Scottish Gaelic, and it is just as likely that the Great Auk was known by an entirely different an Irish name now lost to us.

There are suggestions that the Great Auk might have existed on the north Ulster coast in the past few centuries. Armstrong recounts that when the Reverend J. Guage bought the island of Rathlin in 1740 he left a short paper on the birds that were found there. He mentioned great numbers of birds like Puffins,

G. D'ARCY '85

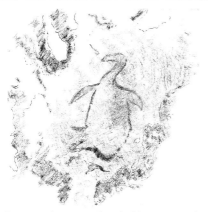

47 *Palaeolithic cave drawing of probable Great Auk, S. France*

Guillemots and sea gulls and also a larger fowl 'bigger than a goose' which he supposed to be a 'penguin'. Dr Pococke in his tour of Ireland in 1752 recorded the Razorbill or Auk 'as big as a pheasant' with a parrot-bill breeding on Horn Head, Co. Donegal. He may well have been referring to the Razorbill (*Alca torda*), but his concept of size is, to say the least, a bit wayward. Sampson (1802) in his *Survey of Londonderry* states that the *Alca Impennis*, Penguin, frequents the rocks of that county and of Donegal. It is impossible now to ascertain whether or not this was first-hand information, but it is hard to believe that such knowledge could be unknown to other 19th-century naturalists. It is most unlikely, for instance, that a large and strange bird like the Great Auk would be overlooked by J.V. Stewart who listed all the birds he encountered in north Donegal from 1825 to 1828 (*Magazine of Natural History*, 1832).

There are a few recent historic references to the bird from Britain. In *Ornithologica* by F.J. Willughby (1676) is the following comment: 'The bird called Penguin by our seamen which seems to be *Hoiers Goifugel* … in bigness it comes near to a tame goose … the wings are very small and seem to be altogether unfit for flight … '

It was mentioned in Wallis' *History of Northumberland* in 1769 and in Blount's *Ancient tenures* in the following extract: 'There is a bird nearly as large as a goose called an Auk, the Alca of Linnaeus, which was allowed at one time to be eaten at Lent.' (The Great Auk was reported as being 'formerly eaten in Lent' when only fish

was permitted by the Church at the time. The exemption was apparently related to their strict pelagic habitat.) There were also historical reports of probable Great Auks from the Isle of Man, the stacks of the Calf of Man being the likely location.

One of the clearest references to breeding came from the isle of Lundy, off the coast of Devon. The report, 'Great Auk on Lundy', (*Zoologist*, vol. 2, 52) mentions that 'During the egging season one of the men brought an enormous egg of the "king and queen murr" [as in 'murre', another name for the Razorbill (Ray, 1674)] described as nesting on a ledge near the water and scuttling into the water on being approached; very few nested there' (Reverend Heaven, 1839).

Our only authenticated historical account of a live Great Auk came from Co. Waterford in May 1834. It was captured (according to Ussher) on cliffs near Brownstown Head and not near Waterford harbour as has often been reported. It was kept alive for four months on fish but subsequently died (see the Note below). The stuffed specimen is preserved in the National Museum. Another, apparently recovered on the strand at Castlefreake in Cork, was not preserved and we rely for veracity on the report alone. Around the mid 19th century there was a spate of Great Auk sightings, most of which can undoubtedly be dismissed as hearsay. However, at least one – two birds together in Belfast Lough on 23 September, 1845 – has a ring of authenticity. It was reported by an experienced wildfowler and was taken seriously by William Thompson.

The story of the decline and extinction of the Great Auk is one of exploitation for food by man. The literature on the bird is voluminous and all too vividly recounted by S. Grieve in his classic book; it is a litany of mindless destruction equivalent to that inflicted on the Passenger Pigeon or the Eskimo Curlew.

A series of adventurer/voyagers – Carthier, Theve, Hore, Parkinson etc. – allude to the practice of killing Great Auks for food and eating their eggs. On Bird Island near Funk Island off the Newfoundland coast, Carthier found huge numbers of Great Auks and Guillemots. He killed over a thousand, filling some thirty boats with them. This was standard practice since these colonies

48 *Sketch of Great Auk from Isle of Man (1651–3)*

were regarded by sailors as self-replenishing 'depots' to provision their boats with fresh meat and eggs. Thomas (1794) provides us with an unforgettable entry in the ship's journal: 'The quantity of birds which resort to this island is beyond belief. As soon as you put your foot ashore you are met with such thousands of them that you cannot find a place to put your feet and they are so lazy that they will not attempt to get out of your way. If you come for their feathers you do not give yourself the trouble of killing them, but lay hold of one and pluck the best of his feathers: you turn the poor penguin adrift to perish at his leisure. This is not a very human method but it is the common practice … Whilst you abide on this island you are in the constant practice of horid cruelties, for you not only skin them alive but you also burn them alive to cook their own bodies with.'

Great Auks were undoubtedly exploited thus throughout their limited range. They declined rapidly as shipping connections consolidated in the north Atlantic. Colonies lingered on, however, on isolated rocky outposts such as St Kilda after they had been eliminated elsewhere. The St Kildans who had 'harvested' the seabirds on their island for centuries would take the auks and their eggs using a birdcatcher, a man suspended from the top of a precipitous cliff on a long rope.

The last colonies were restricted to a few rocky stacks off the south coast of Iceland and would probably have survived longer were it not for the untimely volcanic eruption of 1830 which forced them to occupy rocks which were more accessible from the sea. Armstrong poignantly described their demise: 'In 1844 three men landed and killed the only two birds that could be found. The sum they received for depriving all future generations of the pleasure of studying a quaint and harmless creature was nine pounds.' The sum total of relicts which remind us that this bird ever existed is: about 80 skins; 23 or 24 skeletons; 1000 or so detached bones and 71 or 72 eggs – not much from a bird which existed in countless thousands until only a few centuries ago.

There were, of course, many later reports of 'the last Great Auk' but, with the exception of one which was reported as having been captured on the Faroe Islands in 1870 and thought credible enough to be mentioned in the *Zoologist* (vol. 2, 53), they can be safely dismissed. The main reason for the extinction was of course the uncontrolled exploitation by man. Patton (1906) wrote that 'It seems evident that it was through the active agency of man … that this ill-fated bird was hurried to its doom, and when the birds grew scarce as marketable commodities, it is certain that the last of the species were killed to supply the wants of museums and private collections … '

It should be considered also that other factors may have had a bearing on their decline and demise. We have already seen that volcanic eruption played a part in rendering the relict population on Iceland more vulnerable than hitherto. We do not know now how important a factor this may have been overall: a significant part of their range lies on or close to the geological eruption zone associated with the mid-Atlantic ridge which has shown violent capability (Surtsey in 1963) in more recent times.

Another inescapable factor is that of evolution. The Great Auk exhibited characteristics which suggested that it might be approaching the end of its natural evolutionary development. It was in a sense, an auk anomaly: an outsize, flightless member of the tribe which was forced (by virtue of its evolutionary state) to nest close to the water's edge and was therefore not only vulnerable

to predators (including man) but also to the sea itself. The fact that it laid only one egg meant that it probably could not reproduce fast enough to counter 'natural' depletion.

Ireland was saved the ultimate indignity of the Great Auk's extinction, though people were hunting for it here long after the 1844 record. We will never know the whole story of its former status here, but it seems likely that it was at one time (probably until the beginning of the Early Modern period) both a summer resident in favoured localities around the coast and also a passage migrant from its northern colonies, to winter quarters further south, in regions now unknown.

NOTE

The following details relating to the Great Auk which was captured on the Co. Waterford coast in May 1834 are extracted from R.J. Ussher and R. Warren, *The birds of Ireland* (1900):

He [a David Hardy] referred to the bird spontaneously as a 'Penguin', and said that it appeared to him to have a white ring round its eye (a slight mistake but sufficiently near the truth to show that he remembered this Great Auk). After it had been observed by Hardy swimming about the locality, a fisherman named Kirby captured it without difficulty. It showed so little suspicion that sprats thrown to it inticed it near the boat, when it was taken in a landing net, and appeared to be half-starved. This was stated to have occurred, in May 1834, by Mr Francis Davis of Waterford, who purchased it ten days afterwards and sent it to Mr Jacob Goff of Horetown, Co. Wexford; where it was kept in captivity for four months. For some time it took no food, but then potatoes and milk were forced down its throat, after which it ate voraciously. It was fed chiefly on fish, which were swallowed entire, and trout were preferred to sea-fish. [Goff reported:] 'This auk stood very erect, was a very stately-looking bird, and had a habit of frequently shaking its head in a peculiar manner, more especially when any food was presented to it; thus, if a small trout was held up before it the bird would at once commence shaking its head … It was rather fierce, and seemed to have an aversion to water.' Its subsequent preservation resulted from a request made by the late Captain Spence, when on a visit to Horetown, that if the bird should die it might be presented to Dr Burkitt. This took place and the entry of the fact in Dr Burkitt's collecting-book is as follows: 'September 7th, 1834, I obtained a young Penguin from Francis Davis Esq., which was taken off Ballymacaw, and which I presented to the Museum of Trinity College, Dublin, 1844.'

49 *Silhouettes of Razorbill, Great Auk and Guillemot*

GREAT-SPOTTED WOODPECKER

Dendrocopus major

Despite its name, the Great-spotted Woodpecker is neither very large (9.5 inches; 0.25 m) nor is it very spotted. It is noticeably black and white and has white barring on the wings and sides of the tail which has elongated and stiff central feathers, a functional prop to the bird on vertical tree-trunks. There are also red patches on the head (males and immatures) and under the tail. It is a lively and attractive bird which inhabits woodlands with predominantly mature trees, both deciduous and coniferous.

It is a resident throughout Britain from the extreme north of Scotland to the south of England, being absent only where there is a scarcity of suitable habitat. It is a widespread resident also throughout Europe, being found in every European country besides Iceland and Ireland.

The food (insects, larvae and a range of other invertebrates) is drilled out of the rotten wood and bark by the rapid-fire action of the Woodpecker's hammering. This is a most characteristic woodland sound wherever the Woodpecker is found: people who are not familiar with Woodpeckers often remark about the strangeness of the sound. Great-spotted Woodpeckers are obvious birds, especially when they are feeding young in a nest: they fly to and from the site (a neat hole excavated in the trunk of a tree usually above the height of reach), with a swift, undulating flight and call frequently with an arresting 'chick'.

Although they are more or less sedentary in Britain, Great-spotted Woodpeckers from Scandinavia are (like Waxwings) prone to occasional winter dispersal far beyond their home range. It is thought that the majority of those that have been seen as isolated stragglers in Ireland (mostly in winter) originated in Scandinavia.

Woodpeckers are typical birds of mature woodland throughout the world. They are represented in Britain by three species – the Green, Lesser-spotted and Great-spotted. Three others, the Wryneck, the Nuthatch and the Treecreeper, although 'peckers' of wood and directly associated with woodland, are not, however, true woodpeckers. Remarkably, only one of these, the Treecreeper, is found in Ireland, and there are now no true woodpeckers. In a country which was once among the most wooded in Europe (McCracken et al.) this is at least strange and questionable. An investigation of the evidence available, paltry and fragmented as this is, indicates that this was not always the case and that woodpeckers were once members of our indigenous avifauna.

Great-spotted Woodpecker bones have been found in caves at two localities in Co. Clare. They were found in the more recent, upper layers of calcareous matter on the cave floor and would thus indicate an antiquity approximating to the early Christian period (Mitchell et al.). Bones of Jay and Hawfinch, both woodland species, were found in the same caves, indicating a woodland ecosystem in the vicinity. It would seem that these birds were caught by some now unidentifiable predator and taken to the cave for undisturbed consumption. Evidence of this kind is interesting but inconclusive: more widespread archaeozoological evidence is required for a more coherent picture.

In folklore the woodpecker features prominently in many parts of the world, especially in conjunction with oakwoods. Myths range from the most common – the thunderbird (bird of thunder, storms and rain) – to the imaginative explanation of the woodpecker's red crown, where it was knocked on the head for disobedience. It is strange, therefore, that so little woodpecker fable has survived in Ireland, given that so much has been documented that has an ancient setting. However, when it is considered that a substantial corpus of woodland lore must have disappeared with the final forest clearances of the 17th century, it is perhaps not surprising that the folk memory of the woodland's inhabitants: wolves, deer, noticeable birds also disappeared then.

Irish names have survived and are recorded by Scharff as *Snag*, *Snagdarach*, *Snagbreac* and *Iasair coille*. This last name sounds somewhat speculative as does the name *Gáir chaoilcheann* (literally, 'loud cry of the pointed head') or woodpecker according to Kinsella in his translation of an anonymous 17th-century poem; however, the humorous reference in *Ní Binn do Thorann Lem Thaoibh* (Ugly your uproar at my side): 'Of the two, If I had to choose, I would pick the lesser pain, A woodpecker drilling a tree, Than your snoring in my ear' is highly suggestive of the bird. There is yet another name, *Mórchnagaire*, listed in *Ainmneacha plandai agus ainmhithe* (An Roinn Oideachais, 1978) which sounds something of a concoction in that it is distinguished from the Lesser-spotted (also in the list) only by the prefix *Mór* instead of *Mion*. It seems to be therefore a direct translation from the English.

In Scots Gaelic (according to Scharff) Snag is the Treecreeper whilst *Snagdarach* is the Great-spotted Woodpecker. Nicholaisen, more recently, gives the Scottish name as *Snagan Dariach* and the Irish name as *Snag(aire) Darach*, thus reflecting the close relationship. *Snaicardaraig* is a form used in an Early Irish poem, and this is almost identical to the Scottish name suggested by Scharff. O'Sullivan Beare uses *Snagbruic* or *breac* as in one of the above names which, incidentally, is the modern Irish name for the Magpie. We know, however, that Magpies arrived in Ireland on that oft-quoted occasion in 1676, so this looks like a clear-cut case of name transference. Both the Great-spotted Woodpecker and the Magpie are strikingly pied birds, and it seems reasonable that, with the disappearance of the former and the arrival of the other, the name was simply transferred. The fact that the birds in question were entirely unrelated was no obstacle in the way of a people whose manner of relating creatures to one another was devoid of any scientific premise and based largely on visual criteria.

There are a number of references to woodpeckers in the early literature but they usually lack description and the problem arises as to whether the writer intended an actual woodpecker or a Treecreeper, a bird which has been called 'woodpecker' since at least the early 19th century (O'Sullivan, 1828). In the early medieval Irish poem 'The King and the Hermit' there is the following line concerning the woods: 'There with pied plumage are woodpeckers – in great numbers'. This, on its own, must be taken as a rather fanciful reference – at best indicative rather than factual. More convincing comment comes from O'Sullivan Beare in his *Zoilomastix*. He makes a clear statement as follows: 'The woodpecker is very common in Ireland. A bird of several colours, white black, red, beautiful bluish, with a longish tail, strong beak and hooked nails with which it can hollow out trees and build its nest in the hollow of trees.' While this description is not entirely accurate – unless the bluish is a reference to a sheen – it is good enough to show that the bird intended is indeed a woodpecker and probably the Great-spotted. The author does nothing to expand on his first-hand experience (or our confidence in him) by describing a fable associated with this bird and quoting Pliny as a substantiating reference.

Another interesting fragment is to be found in K'Eogh's *Zoologica Medicalis Hibernia* in which 'woodpecker' is mentioned under the Irish name 'Snagurack' and the lengthy Latin collective 'Merops, Apiastra, Melifuga' in which he combines woodpeckers along with bee-eaters. The Irish name is clearly a phonetic representation (as were most of K'Eogh's Irish names) of *Snag breac*, the Irish/Scots name mentioned above: K'Eogh would presumably have got this name by word of mouth from a knowledgeable local, perhaps a schoolteacher. The temptation is to deduce from this that true woodpeckers were still known in Ireland as late as the early 18th century.

The 19th-century naturalists knew of the Great-spotted Woodpecker only as a rare visitor to Ireland, mainly in winter. John Templeton in his paintings of Irish birds depicted one of these birds, which he stated had been shot in Co. Derry in August

Picus major, a Greater Spotted Woodpecker

nearly the ½ Size

The Specimen from which the figure was taken I received from L.ᵗ McDonnell, It was sent to him from the Co. Derry where it was shot Aug.ᵗ 1802

50 *Great-spotted Woodpecker by John Templeton, c.1800*

1802. Although there is no further comment about the bird in his notes, this is a most unusual season for it to occur, given that most present-day occurrences are as rare winter visitors.

The literature seems to be devoid of additional evidence that might point to the former native status of woodpeckers in Ireland. There is however enigmatic data which broadens the mystery. It relates to the Lesser-spotted Woodpecker, a bird found in the southern half of Britain and throughout most of Europe. None of the modern texts on Irish birds include the Lesser-spotted Woodpecker: it simply is unknown in Ireland nowadays. However, it is specifically mentioned by Smith in his *Antient and present state of Waterford* (1756) and Rutty in his *Essay towards the natural history of Co. Dublin* (1772) and by Thompson, Watters and Ussher at various times throughout the 19th-century.

It is worth reviewing the references in the light of the overall context of the former existence of woodpeckers here. Rutty does not describe the bird other than to mention its Latin name, *Picus varius minor*, that it 'feeds on insects under the bark of trees'. Watters uses Smith and Rutty to argue for its inclusion (to obvious disagreement from Ussher) as a former indigenous Irish bird. It should be stated, in defence of Ussher's criticism, that Rutty does not include the Treecreeper in his list of some 140 birds and who is now to say that this is indeed his 'Lesser-spotted Woodpecker': the Treecreeper could be described as both 'lesser' and 'spotted'. In the Ordnance Survey Memoirs of Co. Antrim (1831–2) the name 'woodpecker' is mentioned in a list of birds encountered on or near Lough Neagh. As the Treecreeper is again not included in the list of 66 birds it is doubtless the Treecreeper which is intended. In general, the name 'woodpecker' is still widely used as a country name for the Treecreeper.

There is, however, a more compelling reference to the Lesser-spotted Woodpecker. It comes from the appendix in William Thompson's *Natural history of Ireland* (1851). One of Thompson's many correspondents, a certain taxidermist, Mr Glennon ('of Dublin') stated that 'In the course of many years he had preserved at least six or seven of these birds, sent to him from various parts of Ireland.' This same Mr Glennon had reported to Thompson the receipt of several Great-spotted Woodpeckers (including one in May), so he obviously knew the difference between the species.

Watters' commentary in his *Natural history of Ireland* adds: 'Of these specimens (the Lesser-spotted Woodpeckers seen by Glennon) the two last we had the pleasure of examining – one an adult male obtained on the 21st September, 1848, in the county of Wicklow; and the other, an immature female, shot in the same county during the autumn of 1847 … '

Whatever we might say about Watters' knowledge of Natural History, and he certainly borrowed heavily from Thompson, his information in this case is unequivocal and in my view sound. The reason for not including the Lesser-spotted Woodpecker in the formal list of Irish birds is obviously due to the loss of the Glennon specimens.

The same is true of Green Woodpeckers which Thompson, Watters and Ussher mention from two or three specimens shot in Ireland in or around the 1840s. The loss of the specimens seems to have been the critical factor in the exclusion of the bird from the Irish list.

Today the status of the Great-spotted Woodpecker in Ireland has hardly changed. It remains a rare visitor, one or two turning up in most winters. Examination of the skins of dead specimens show that they are mainly of the northern race (from Scandinavia) rather than from Britain. Deane (1954) records that the Great-spotted Woodpecker has once or twice turned up in spring and on one occasion two were seen together in the Glenshesk valley, Co. Antrim, for almost two weeks. The possibility of eventual recolonisation cannot therefore be ruled out despite the fact that the area of suitable hardwood habitat remains pitifully low by European standards.

In Scotland the Great-spotted Woodpecker has made a widespread recovery, having more or less disappeared by the mid 19th century. It was well known throughout Scotland up to the 17th and 18th centuries and declined as a result of a number of factors, one of which was stated to be the extensive tree-felling and coppicing, characteristic of that period in Scotland. It has been documented that, as in Ireland, a number of woodland bird and animal species was eliminated or reduced to relict status as a consequence of this exploitation. In David Lack's famous monograph on the Swift there is an interesting aside concerning the Great-spotted Woodpecker. He mentions that about 1835 the Swift was known to utilise old woodpecker holes in the ancient pine-forest along the river Beauly, in Ross-shire. 'What remained of the old Highland forest was later destroyed and the Great-spotted Woodpecker became extinct in Scotland so this habit [nesting in woodpecker holes] must have died out.

The loss of this (and perhaps other) woodpeckers from Ireland seems to have coincided with the final woodland clearances in the 17th and 18th centuries and must surely have occurred as a consequence of this activity. The few references from the 18th century suggest that they may not have been as common and widespread throughout Ireland then as O'Sullivan Beare implies. It should be remembered though that he was from south-west Munster which remained heavily wooded later than many other parts of the country. So his familiarity may have been local rather than regional.

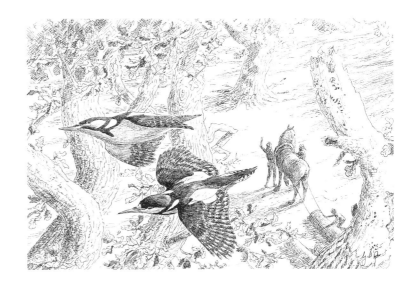

51 *Pair of Great-spotted Woodpeckers in flight*

It is possible, given the scattering of 19th century records of 'obtained' specimens and the inexplicable cluster of Lesser-spotted and Green Woodpecker specimens, that these were the last few native wild birds 'cleaned up' by specimen collectors. This might appear far-fetched to modern birdwatchers – who would regard the likelihood of seeing a woodpecker wild in Irish woods as being of the same order as seeing a Capercaillie – since there have been no authenticated sightings of either Lesser-spotted or Green Woodpeckers in the present century. This lack of sightings substantiates the sedentary, non-migratory nature of these birds, further pointing to the relict probability of the Glennon specimens.

The question of other avian losses arises at this point. It is interesting, for instance, to read Thompson on the Jay, a bird with which he was not familiar despite the fact that he had been told of its former existence in oak woodlands near Lough Neagh that had 'since fallen to the axe'. He refers to the clearances of the woods as being responsible also for the decline and disappearance of certain birds of prey. Were Redstarts, Wood Warblers and perhaps Pied Flycatchers once widespread and common in Irish woods? Nowadays they cling on only in one or two oak woods in the east of the country. Perhaps the increase in interest in and improved methods of identification of bird bones will help us reconstruct the picture of the lost avifauna of our former woodlands.

Sleeman (1993) has posited the stoat as a possible contributor to the demise of the tree-hole nesting birds (Redstarts, Pied Flycatchers and woodpeckers) here. Given the proven arboreal habits of the Irish stoat and our relative scarcity of mammalian prey (compared with Britain) this theory is certainly worthy of consideration. However, it is likely to have been significant, if at all, only in the latter stages of decline, when suitable trees were at a premium.

Some interesting work has been carried out to investigate the fossil tree trunks preserved in our peat bogs. Entomologists have been concentrating on the remains of *Pinus sylvestris* in tracing the continuity of a range of invertebrates which continue to exist here despite the loss of their original natural conifer habitat. Some species have disappeared completely over the thousands of years. One of the looked-for signs are the tell-tale holes made by woodpeckers – both the drill holes for food and the neat nest-holes. As yet no clear-cut evidence has been forthcoming, but the number of trunks examined in comparison with stumps is small and the rotten parts of the pine favoured by woodpeckers would be the first to disintegrate in the bogs. This, nevertheless, remains a hopeful avenue of investigation. Undoubtedly time will help us to develop a clearer picture of the lost woodpeckers.

The possibility of the Great-spotted Woodpecker repopulating the increasingly available woodlands increases with every year. It is not nearly so sedentary as the others and occurs regularly, usually as a single bird. As mentioned, the majority appear to emanate from Scandinavia – birds erupted in a similar fashion to Waxwings or Crossbills. However, there have been occurrences in spring – one at Ballycarry, Co. Antrim, on 5 May 1950, and the other pair which turned up at Glenshesk in the Glens of Antrim in April 1931 and were thought by the observer to be the British rather than the Scandinavian race (they are said to be indistinguishable in the field) and were described as such in the report of their occurrence in the *Irish Naturalists Journal* of that year. One shot in Co. Fermanagh on 12 December 1959 was examined in the hand and found to be definitely of the British race (Hutchinson, 1989). The Antrim birds did not apparently stay to nest, but their arrival together (assuming that they were a mated pair) into suitable breeding habitat indicated their potential to do so.

52 *Woodpecker drill-holes in rotten wood*

PART THREE

TOMORROW'S LOST BIRDS?

The memory of birds that 'used to be', whether accurately recalled or perceived, seems often to be a part of the package of growing older. How many times have we listened respectfully to the tale of the meadow which half a century ago was full of Skylarks and now hasn't even a grasshopper. Birds are the oft-called upon symbols of the sunny, song-filled days of youth. Perception, particularly in retrospect, is often as important as fact.

While conscious of this it is nevertheless necessary to point to the ongoing potential for loss among our indigenous bird population. That a number of our birds teeter on the brink of disappearance is an undeniable fact and it serves no purpose to ignore this for the sake of steering this book towards an optimistic conclusion. In fact the exploration of the factors contributing to these declines must be one of our primary concerns.

A number of birds (the Red-necked Phalarope, for instance) which inhabit Ireland at the extreme edge of their range and have probably never been widespread, are certainly teetering. For as long as this delightful little Arctic bird has been known to nest here (since the early 1900s) it has been confined to one or two far western localities and looks now like it will never become established. But there are other wading birds, which are declining alarmingly, having been widespread inhabitants of certain freshwater wetlands – callows, water-meadows, wet pasture – habitats which have come under increasing pressure in recent decades due to land reclamation. The Dunlin is a case in point. But most of the wading birds which breed in wetland – the Snipe, the Curlew, the Redshank, and particularly the Lapwing – are not as common as they used to be. There are strongholds, of course, such as along the callows of the river Shannon between Loughs Ree and Derg, a habitat which significantly avoided arterial drainage. It is interesting too that a newcomer, the Black-tailed Godwit, may have successfully nested here in recent years amid the other thriving wader populations. Thankfully there is enough conservation focus in this region to provide protection for this otherwise threatened suite of birds.

Another group of birds which is under threat is the game birds – birds which have been traditionally hunted as quarry species. We have already highlighted the Capercaillie, which was certainly avidly hunted for food, though as the case study pointed out, this was only part of the story. But our other game birds are also in a sorry state. We may ignore the Corncrake (which was also a traditional quarry species) for now, for, like the Capercaillie, it will be considered separately. The Pheasant, the game bird with which we are most

53 *Red-necked Phalaropes*

familiar is, of course an introduced species from Asia brought here for food and sport. It is nowadays widespread and common in most parts of the country and in no imminent danger of decline. It is on three others – the Partridge, the Quail, and the Red Grouse, which as far as we know are indigenous – that we must focus.

To date we are lacking in conclusive fossil evidence concerning this trio of gamebirds, so it is difficult to trace their background as natives. However, the existence of Irish names for each – *Gearra guirt* for Quail, *Patraisc* for Partridge and *Cearc fraoigh* for Red Grouse – points in each case to ancient lineage. The Quail echoes the Corncrake in being a summer visitor to Ireland, while the Partridge and the Grouse are non-migratory or sedentary. Like the Corncrake, the Quail has for millennia been trapped for food in the Mediterranean countries on passage to African winter quarters. There can be little doubt that the abundance of the Quail down the ages was closely related to the conditions which permitted the flourishing of cereal crops. It is plausible that, like the Corncrake, the Quail has been identified with the waxing and waning of agrarian activities ever since cereals were first cultivated in Neolithic times. There are references to Irish Quails from Tudor times and there may be other, much earlier, ones too, though it does not seem to be specifically mentioned in the early Christian material. Partridges are mentioned in a 14th-century Norman reference to hunting (with hawks) in tilled land in the vicinity of Dundalk. There are also clear allusions to both Quail and Partridge from Tudor times. However, the richest store of historical reference comes from H. O'Sullivan's diaries (*c*.1828). The author regularly comments on the sounds made by the game birds in the cereal fields around Callan: 'I hear the Quail crying *fuid fuide* and the Corncrake crying *aic aic* ... ' Also, 'Fowlers shoot Quails in winter and I hear them summer and autumn. Consequently the Quail does not leave us nor is it a migratory bird of passage as I thought, or perhaps it is the strong ones that go and the weak ones stay.'

Thompson's *Natural history of Ireland* (*c*.1850) provides us with another benchmark in the record of our game birds. While all were still abundant and widespread, both the Quail and the Partridge had retreated somewhat, even then. He expressed concern about the Partridge, which he stated had, to his knowledge, never been as common as it was in Britain. This relative scarcity is born out by Fynes Moryson more than 250 years earlier. Ussher and Warren, who witnessed the marked decline, suggested that it was due to 'the discontinuance of wheat-growing and the use of breech-loaders combined with poaching' and also, significantly, that the wetness of Ireland was less favourable to their increase than in Britain. They mentioned the lack of wintering Quails as indicative of the noticeable decline that was underway around 1900. Moffat (1896) also noted the striking decline and hinted at a correlation between the wetness of the climate and the abundance or otherwise of the Quail.

From having been a widespread resident in the late 19th century, the Quail had become confined to a few east-coast localities by the later 20th century. One of the years covered by the *Atlas* (1970), was described as a 'Quail year'. Even so, its breeding success was more or less restricted to those traditional grain-growing counties, Meath, Wexford and Kildare and a few outliers.

The *Atlas* results for the Partridge indicated that in the years 1968–72 it still bred in most Irish counties; however, the results were prophetically described by the editor as 'deceptive': 'The Irish

54 *Quail*

decline was so serious that protective legislation prohibiting or restricting shooting of Grey Partridges was introduced in 1930 [1931 in the Republic]. Without this and restocking, the species would probably have become extinct there [in Ireland]. Even now it is very sparsely distributed and scarcer than the map suggests … ' (Sharrock, 1976).

Over the past quarter century the decline has continued. Less than 80 pairs were counted in 1995, the majority in Co. Wexford: an investigation indicated that the last few (less than 30) remained in a tract of abandoned cutaway on the edge of one or two raised bogs in Co. Offaly (*Irish Birds*, 1997). Now at the end of the century it is debatable whether enough remain in the wild to enable increase (under conservation conditions) to occur.

The Quail too remains only as a sporadic summer resident with pockets in certain traditional strongholds mainly in the wheat and barley districts of the east. One surprising development which gives hope for the future survival of the species is its occurrence in the callows of the Little Brosna on the borders of Offaly and Tipperary. Several have been heard holding territory (and presumably breeding) here in recent years, and, in view of the conservation programme now under way for the Corncrake, the Quail (and the Spotted Crake) may be unexpected beneficiaries.

The fate of our other game bird, the Red Grouse, is by no means assured, though it is presently not as precarious as the others. Its 20th-century decrease has nevertheless caused considerable alarm. Its pedigree as an Irish endemic is without question. Its unambiguous Irish name, *Cearc fhraoigh* (heather hen), has been mentioned in the early literature and in the Celtic poetry: it has undoubtedly ranked as a favourite bird for the pot, throughout human history in Ireland.

The 19th-century writers regarded the Red Grouse as a widespread and abundant gamebird, 'maintaining its ground successfully against all the engines of slaughter directed against it' (Watters, 1853). But Thompson noticed that its numbers were declining from 'Various causes connected with the operations of man' and that it 'was not protected with the same anxious care as in Scotland'. Ussher and Warren (1900) restated this but said that

it still bred in every county. Ruttledge (1954) stated that there had been a marked decrease in the 20th century, particularly latterly. Conservation in Ulster, he said, had been more effective than in the Republic, where the activities of Bord na Móna had begun in earnest in 1946. The *Atlas* (1976) showed clearly that the Red Grouse had indeed decreased in range, though it still bred sparsely in most Irish counties: it had by then vacated many of its Midlands raised bogs.

Investigation into a concurrent (but smaller) decline in Britain indicated that habitat degradation was a factor. It was found, for instance, that to maintain a viable stock of Red Grouse on a particular moor, the habitat needed to be carefully managed. Heather burning, predator control etc. are part of this procedure, but when this occurs indiscriminately or in spring when the grouse are nesting (as it does in many parts of Ireland), the stock suffers. The highly territorial males may be driven out and the population decimated as a consequence. However, it is possible to make too much of this. Heather burning has occurred naturally since time immemorial – from lightning in especially dry periods – and our bogland ecosystems are subject to less predator attention than was formerly the case. The oft-spoken adage applies: Leave the habitat (with access to others) and the wildlife will look after itself. In any case the Grouse density here is unlikely ever to have been as high as that in the luxuriously covered heathlands of the Scottish highlands.

However, the Red Grouse is habitat-specific: it needs heather. Thus peat-cutting and afforestation have reduced the terrain available. There can be no denying the impact of the past half-century of bog exploitation: cut-away bogs (even where these support new growth of heathers in the plant matrix) are no substitute for the pristine heather-clad peatland.

Restocking of the Irish population of Red Grouse has been going on for many years. This has helped to maintain stocks for hunting, particularly in the Ulster moorlands, but it has unfortunate ramifications. The Irish Red Grouse is a distinct race, quite separate from the Scottish race. Along with the Dipper, the Jay, and the Coal Tit it belongs to a select group of endemics –

subtly distinct from their British counterparts. By hybridising with introduced stock, this endemicity with its potential for genetic and other study (having developed imperceptibly over thousands of years) is occluded. The counter argument of course is: Do we stand idly by while a population of birds – endemic or otherwise – dwindles into oblivion for the want of a 'helping hand'?

The efforts of the Irish Peat Conservation Council in setting aside a small number of unexploited raised bogs in the Midlands may go a long way towards providing a most effective helping hand that might prove to be the difference between extinction and survival of Ireland's own unique Red Grouse.

Another group of birds which have shown disturbing signs of decline are those associated with old-fashioned agriculture, unfenced lowland and upland commonages and coastal regions; in general, uncultivated or abandoned corners of the countryside. They may be collectively referred to as 'Rough-country birds'. These birds were in their element from the time of the first woodland clearances (for they would probably have been rare or local in the unmodified landscape) right up until the decline of mixed farming – the advent of land reclamation by machinery, crop sprays, monocultures, genetically controlled planting etc.; in short, the coming of agricultural intensification within the last half-century.

55 *Red Grouse and peat-burning power station*

Both the Corncrake and the Grey Partridge are associated with this group, but it also contains a number of birds which, even up to thirty years ago, were regarded as common and widespread. The list is small – the Nightjar, Ring Ouzel, Corn Bunting, Chough, and Twite – but unfortunately may grow (the Yellowhammer?) as agriculture continues on its intensive lunge. It is interesting how these birds echo the decline and (in some cases disappearance) of a group of plants – victims also to modern agriculture. They include the 'ruderals' – Darnel, Fat Hen, Corn Cockle, Cornflower, Corn Marigold etc. Were it not for the recent discovery of Corn Cockle and Darnel on the Aran Islands, these plants – at one time as commonplace as the crops in which they grew – would have passed from our consciousness without so much as a whimper.

There is very little mention of the 'Rough-country birds' from the early writers. Those who considered birds were drawn to the larger of showier species. Historical material of any antiquity is thus lacking. A complication lies in the fact that the English names of smaller unobtrusive birds have often been interchanged, misapplied or simply lacking. The Irish is often no more helpful, and references earlier than the Early Modern period are few and far between. It is possible that there is some mention in Early Irish poetry, but to date (due to difficulty in deciphering) their identity remains obscure.

The only name that can be found for the Nightjar in Irish is *Túirne lín*, a descriptive name for the bird's highly distinctive churring song. It means literally 'flax spinning wheel'. The name is probably as old as the activity. However, in keeping with its mysterious nature there seems to be very little known about it prior to the 18th century. It is not mentioned by Rutty, Harris or Smith, but Templeton knew of it in Ulster at the end of the 18th century. To Thompson it was well known as 'a regular visitor to favourite localities in all quarters of the island [Ireland] … '

Ussher and Warren provided a much more complete picture of the Nightjar's status *c.*1900. They knew of it nesting in no less than twenty-five counties and perhaps others besides. A series of regular breeding haunts is mentioned for the eastern side of the country, from the Clandeboy estate and Mourne mountains to Howth

Head, Lambay, Carrickgolligan and the Dublin mountains, to Tara Hill, Avoca, Co. Wicklow and Forth Mountain, Co. Wexford. By 1985 only Forth continued to be occupied by Nightjars and it may have been vacated since then. The 1976 *Atlas* showed that the Nightjar was indeed a thinly distributed and very local breeding bird in Ireland: 'Compared with numbers at the end of the 19th century, when this was considered to be a common bird, there has been a drastic decline.' The abandonment of the East coast sites has obviously been replicated throughout the country, for when the *Red Data Book* of endangered vertebrates was compiled (Whilde, 1993) the total estimated breeding population was thought to be as few as 30 pairs. While one or two previously unrecorded sites have been discovered, it is difficult to see how at the present level of breeding the Nightjar can continue to survive here.

A corresponding decline has occurred in Britain but the situation is nowhere near as drastic. Sharrock, editor of the *Atlas,* wrote about the Nightjar as follows: 'We still do not know the ultimate factors responsible for this long-term decline, which has been shared by certain other insectivorous summer migrants … the felling of woods, the increased building on heaths and commons, the disturbance by trippers at weekends and during holidays may have contributed … Nor is there any obvious correlation with climatic change, which might influence the level of insect food for Nightjars … [which] were declining during the warmer period of climatic amelioration which was so marked up to the 1939–45 war. Pesticides may have been a contributing factor in the 1950s and 60s, but these cannot have done more than exacerbate a situation which already existed.'

Smith (mid 18th-century) referred to the 'Rock Ouzel' – an obvious reference to the Ring Ouzel – and stated that its Irish name was *Lonyleac* (literally, 'rock blackbird'). Older references are obscure or non-existent, though *Lon dhubh* for Blackbird can be traced back to the early Christian poetry. Smith knew of this bird from the Mourne mountains and the Cooleys and abundantly from Mount Leinster in Co. Carlow. Thompson (1848) stated: 'Is found during summer in suitable localities over the island

56 *Ring Ouzel pair*

[Ireland]', and he described its known haunts in some detail. It must have been abundant in places, for he described how it could be seen in flocks in autumn, even on the high ground to the west of Belfast.

When Ussher and Warren considered its status in 1900 they reckoned it to breed in twenty-seven out of the thirty-two counties. By 1966 Ruttledge found that it had declined very considerably and that its breeding distribution was local and scattered. The *Atlas* mentioned that, allowing for the possibility of overlooked breeding sites, there had been an alarming decline. It was recorded as nesting or probably nesting in only a handful of counties – mainly in the Leitrim/Sligo and Wicklow mountains and (in common with the Nightjar) represented by only a few dozen pairs. Subsequent investigation suggests that it also continues to occupy upland regions in Counties Down, Kerry, probably Cork and Donegal, so it is not as imminently endangered as the Nightjar.

In Britain, where the Ring Ouzel is much commoner and is not restricted to the mountains, there has been a substantial decline also. This has been attributed to the spread of the Blackbird and perhaps the Mistle Thrush rather than direct, man-centred activities. Williamson (1975) was of the opinion that climatic change may have had a bearing, causing the Ring Ouzel to retreat further into the uplands. In its Irish range this bird comes less into contact with man than in Britain, preferring mainly inaccessible escarpments usually above 500 metres. However, with agricultural intensification (including high sheep densities and ubiquitous forestry) these former nether regions are now subject to increased disturbance.

How long the Corn Bunting has been an Irish bird is unknown but, as in the case of the Corncrake, it surely goes back at least to the early days of cereal growing and beyond. An Irish name is *Gealbhan an ghuib ramhair*, translated by Ussher and Warren as 'Sparrow of the thick bill', a most satisfactory descriptive name for this bird. It is unfortunately impossible to trace the name into antiquity as *Gealbhan* has been used for centuries as a generic/collective name for birds of the finch or sparrow family.

Undoubtedly the rather featureless plumage of this bunting has caused it to be ignored: it was just one of those little brown birds which lived, generation after generation, in close proximity to rural communities. It was overlooked in many lists of birds even into the 19th century, despite being a familiar, widespread resident. It was listed as common by O'Sullivan in early 19th-century

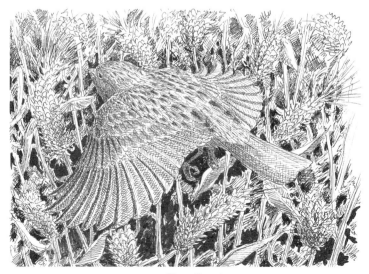

57 *Corn Bunting*

Kilkenny and by Thompson as a permanent resident, throughout the country in the mid 19th century. Thompson's name for it was the 'Common Bunting' and the other names, 'Briar Bunting' and 'Corn Bunting' were names used locally in the north of Ireland. In the south it was locally called the 'Corn Bird'. In those days, in hard winter weather, Corn Buntings gathered (like sparrows today) in less frequented streets and stableyards in Belfast. Others, notably Watters and Brunton, mention the common status of the Corn Bunting in the 19th century. Ussher and Warren, as usual, provide us with a comprehensive coverage of the bird's Irish range. They stated that 'It was generally common in coastal districts, local elsewhere' and that it was then (1900) breeding in every Irish county bar Leitrim and with doubt as regards Fermanagh. In a charming section about local names they mentioned that a 'peasant youth' named Corn Buntings 'Tittery hay-birds', a name full of resonance both for the 'bunch-of-keys' song and the crofter-style landscape which it inhabited.

When Ruttledge wrote on Ireland's birds in 1966 he stated that the Corn Bunting had disappeared by then from its inland haunts and was a scarce and local bird in some coastal haunts. Deane (1954) corroborated this rapid decline in the north. When the *Atlas* was published in 1976, the range was restricted to the coastal fringe of a few counties notably Down, Wexford, Waterford, Kerry, Galway, Mayo and Donegal. The 'crash' of the Corn Bunting in the course of half a century puzzled many ornithologists: the reasons were not at all clear especially in the west where ideal habitat was abundant: 'It is doubtful if the Corn Bunting could have reached much of central and western Europe if it had not been for man's influence in creating the agricultural landscape on which it has become so dependent. It seems strange, then, that having adapted to these circumstances, presumably for several thousands of years, it would surrender them so abruptly in Ireland – unless some other factor associated with its lifestyle had impinged upon it.'

In Britain, where the population of Corn Buntings coincided with the coastal agricultural belts, there has been a similar decline, particularly in Scotland and Wales. The factors influencing the decline in Britain were (in 1967) stated to be the mechanisation of harvesting and perhaps some climatic considerations. This was also thought to have contributed to the general decline over most of Europe. Significantly, where agriculture continues to be extensive and traditional (in eastern Europe), the Corn Bunting continues to thrive. The *Atlas of European breeding birds* (1997) pointed to the unavailability of winter stubble due to agricultural intensification and advanced harvesting techniques as being the major pressures affecting this and other farmland birds. Taken in conjunction with climatic pressures which have most effect on bird populations on Europe's western fringe, these are seen to be the key factors in the western decline.

These could also be the factors adversely affecting the Yellowhammer here. A general decline has also been noticed in recent years, but it is in no way as significant as that which has removed its dowdier relative. In 1998 no Corn Buntings at all could be found in the well-known sites and it is feared (1999) that it may have become entirely extinct in Ireland.

Two other birds, the Twite and the Chough, are worthy of consideration in this section though they are by no means as threatened as the foregoing. They have both declined and occupy ranges more or less restricted to the western seaboard. As is the case with the Corn Bunting and other small nondescript birds, the Twite was undoubtedly more familiar to country people in the past, though it has not been distinguished in the literature. Templeton and Thompson mention it, but it is clear from their brief references that they knew little of its lifestyle. Nonetheless it was a familiar bird to grouse shooters and others who frequented the uplands, where it was called the 'Heather grey' to distinguish it from the Linnet or 'the grey'. In Thompson's opinion it was then widespread in the mountainous regions in both the north and the south of the Island. It was known from the Ulster mountains and from coastal areas in winter where it gathered (as today) in flocks. At the end of the 19th century it was known to breed in a number of places – around Lough Neagh, in the Dublin and Wicklow mountains, in Waterford – where it is not found today, and it is evident that its Irish range has contracted in the interim.

58 *Twite pair*

Now there are reckoned to be less than a thousand pairs, confined to the west with strongholds in Kerry and Donegal. The European *Atlas* cites overgrazing by sheep and the introduction of modern agricultural techniques into formerly traditional areas as leading to the decline. However, there may be geographical and climatic considerations too. The world distribution of this finch is most unusual. The bulk is found on the Asian steppes. The western outlier – a relict population, isolated due to glacial separation during the Ice Age – is confined to a western European fringe and includes the Irish birds. It could well be therefore that the climatic and geographic factors which affect most species particularly on the edge of their range are also weakening the Twite in Ireland.

The charm of the Chough has been noted by many in Ireland down the ages. Its Irish name *Cág cosdearg* (Red-legged Jackdaw) is often mentioned and it is well known for being used as a symbol on the medieval heraldic emblem of the O'Kirwans, one of the tribes of Galway.

O'Flagherty and Molyneux recorded inland breeding sites of the Chough in the west of Ireland and was known from other inland nesting sites by 18th- and 19th-century writers. Thompson recorded also that it nested at several places on the east coast.

Though Ussher and Warren knew of it from the coasts of all the provinces and in a few inland locations besides, the Chough was already a 'diminishing species' in 1900. A gradual decline and shrinking of the range occurred over the first quarter of the present century and, though the population has stabilised since, it is still common only on the grassy-topped rocky coast and machair of the western seaboard.

'The maintenance of sheep-grazing pastures on undisturbed soils is considered critical for the future of the Chough in Ireland' (Bullock, Drewett and Mickleburgh, 1983). Recent censuses have shown that the population has indeed stabilised at around 750 pairs, over a quarter of the western European population. But when one considers that there has been a noticeable shrinkage of its range over the whole of Europe, there are obviously no grounds for complacency.

Currently anxiety has been expressed about declines in some other species of birds and while most bird populations are subject to periodic fluctuations it is the long-term downward trend that must concern us. The Barn Owl is a case in point. Though censusing has revealed it to be commoner than was thought, it is still much reduced compared with, for instance, Ussher's time. The

reasons for the decline are unclear but could be related to the general change from grain to pasture resulting in a reduction in the availability of rodents (the owl's main food), destruction and reconstruction of old buildings (traditional nest-sites), pesticide overuse, hunting and accident etc. To date the situation is not critical and considerable effort has been expended in providing artificial nest-boxes and in otherwise providing increased protection for our remaining Barn Owls. It is too early as yet to determine whether the population which is highly sedentary has stabilised.

We must reluctantly return again to our beleaguered birds of prey. Of our six remaining species, the Peregrine, Kestrel and the Sparrowhawk appear to be doing best (though this was certainly not the case a quarter of a century ago when organochlorine seed dressing was taking its toll on the predators at the top of the food chain). Two, the Merlin and the Hen Harrier, have been dwindling in numbers of breeding pairs. The *Red Data Book* (Whilde, 1993) leaves us in no doubt. The figures mentioned – slightly more than a hundred pairs in the case of the Merlin; less than a hundred pairs in the Hen Harrier's case is disturbingly low.

Both are moorland nesters and have been affected by similar circumstances – the gradual exploitation of their habitat and particularly forestry. Paradoxically, the increasing spread of forestry plantation has disadvantaged and aided both species. The disadvantage has been as stated, but both Merlins and Hen Harriers have also adapted to utilise forestry plantations for breeding – the Harriers when the trees are very young, the Merlins when they are mature and lie on the edge of open moorland. However, these benefits aside, both birds continue to decline, and the uniform maturing of Sitka Spruce plantation presents a hostile environment to the more endangered Hen Harrier. A 1998 survey undertaken by the Irish Raptor Study Group, BirdWatch Ireland, and the National Parks and Wildlife Service gives little cause for optimism. There has been little change in five years – 58 definite pairs and 13 possibles, in the Republic. A parallel survey in the North found 30–35 pairs. The conservative estimate of less than 100 pairs on the island still seems to hold. This contrasts markedly with the 200–300 pairs of the early 1970s. Hen Harriers seem

59 *Merlin and meadow pipit*

reluctant to move from traditional breeding grounds to exploit new areas, many of which have subsequently been reclaimed or rendered unsuitable otherwise (Norris, 1999). They have, however, shown a capacity to occupy lowland cut-away bog and even fenland. In the absence of the Marsh Harrier, traditional denizen of such terrain, wetland might offer an alternative to those Hen Harriers able to adapt to new circumstances. It would be a major ecological tragedy if we were to lose our only other harrier to modern habitat modification. The Merlin does not seem quite so threatened. Being much smaller and faster on the wing, it is also much less likely to be shot than the large, dallying harrier. While tracts of blanket bog such as those at Roundstone or in Donegal remain with their abundant populations of Meadow Pipits and Skylarks, it would seem that the Merlin can survive.

Finally we must consider two of our five species of terns – the Roseate and the Little Tern – which are ranked respectively as 'endangered' and 'vulnerable' (Whilde, 1993) due to their precariously small breeding populations. Recent counts indicate c.650 pairs for the former and c.300 for the latter. Information about their status in historic times is hard to come by and there is no repository of bone evidence to call upon. It is thought that the Roseate, at any rate, was abundant in a few east coast localities in the first half of the 19th century. However, it is possible that neither bird was particularly common in the past. Their small population could be as much a factor of geographical location – at the edge of their European range – as anything. Their numbers have been subject to wide fluctuations and colony locations are constantly changing. This has been put down to climatic change – warmer summer conditions in the first half of the century; cooler summer conditions since (Reid-Henry &

60 *Hen Harrier pair*

Harrison, 1988). Interestingly, a litany of other threats is cited, including: predation by gulls and Hooded Crows and by rats, foxes, stoats and other mammals; disturbance of the nest sites (some of which are on resort beaches) by humans; flooding by spring tides; removal of protective vegetation. In addition, many Roseate Terns are killed for food (and sport) by local people at their winter quarters in West Africa, highlighting a common problem in protecting a migratory species.

Thus our Roseate and Little Terns, in the complexity of their circumstances, exemplify the difficulty facing our new millennium conservationists, as who strive to protect our threatened avifauna.

THE CORNCRAKE PROFILE

The Corncrake or *Traonach* is a small, chicken-like ground bird of the same family as the moorhen. Its biscuit-coloured plumage is liberally and generally streaked and spotted, and it is greyish on the breast and face. There is a distinctive rusty patch on each forewing.

It is rarely well-seen, preferring to remain hidden from view in high grass etc., from which it is extremely difficult to flush.

Its presence is signalled by its unmistakable call (really the song of the male bird), a raucous double-note *crek crek* repeated for minutes or even hours at a time, often through the hours of darkness. Those who hear it for the first time often remark on its mechanical quality. Indeed, it can be accurately simulated by rasping two serrated bones together.

The Corncrake is migratory, spending the winter in Africa, mainly in the south and south-east of the continent. The eastern route around, rather than across the Mediterranean, is preferred by many (probably the bulk population of eastern European birds), while a different route – a more likely one for western European birds – through Algeria and Morocco is also recognised (Batten et al., 1990). The Corncrake arrives in Ireland towards the end of April at about the time of the Cuckoo. Initially the favoured habitat is damp, rushy pasture: eventually they move into drier areas – hay meadows and rough grassland.

The nest is characteristically in a grass tussock, and a large clutch – from six to twelve eggs – is laid in May. By early June the fluffy black chicks are out and are being fed on invertebrates by the adults. Two broods are thought to be normal, even without the loss of the first. By September the young are fully-fledged and the adults have moulted in preparation for migration. Before mid October they have left on their 3,000–5,000 mile flight – across Europe, the Mediterranean, the Sahara desert and a good deal of equatorial Africa – to their savannah or veldt winter quarters. For birds such as Swallows and Cuckoos the flight is a remarkable feat: for the awkward Corncrake which takes to the air reluctantly and dangles its feet as though anxious to return, this annual migration verges on the unbelievable. As if to broadcast this remarkable capability, the Corncrake has turned up in places as far afield as Australia, New Zealand, North America and Greenland, thousands of miles away from its normal range, which extends across the temperate zone from western Europe to central Asia.

Due to the decline in its population, more or less throughout its range (but particularly throughout western Europe), the Corncrake is regarded as being globally threatened; its present status in Ireland renders it imminently threatened with extinction here.

There is sad irony in the fact that the last stronghold of the Corncrake in Ireland is in the meadows of the callows of the Shannon, given that perhaps the earliest literary reference is from the same place: '*Dá traghna a srothaibh Sinna*' (Two corncrakes from the banks of the Shannon), part of the ransom to be paid to Cormac Mac Airt for the release of Fionn Mac Cumhaill (from a 9th-century poem, trans. E. O'Curry). Another early extract which could be a reference to the Corncrake today lists it along with nettles and elder bushes as one of three signs of abandonment (Old Irish triad).

An Traonach has been extolled in literature many times since, its singular voice having captured the imagination of both bard and

chronicler down the ages. The 9th-century poem, 'May Day', in listing typical elements of the early Irish countryside refers to the 'Corncrake chorus'. Lady Gregory in her translation of Finn's song referred to the Corncrake as 'loud-voiced poet' (from *Gods and fighting men*). More recently Richard Murphy saw it as a 'crepuscular, archaic politician', and Ledwidge, wandering out in a rainy April night, noticed how the 'pilgrim moon' plied the sky to the 'lonesome tune of the brown meadow rail'.

The wry anthropomorphism of the early writers suggests a tolerant affection for the bird. It is frequently compared to people in the texts. In one example its repetitive call is compared to a snoring husband; in another, a nagging wife or some other annoying individual.

A testimony to its familiarity from the earliest times lies in the enshrinement of its Irish name in placenames. P.W. Joyce pointed out that the Irish name was pronounced '*tryna*' in the south and west but '*traina*' elsewhere and is anglicised in Coolatreane (the meadow of the Corncrake) in Fermanagh, or Lugtryna (the hollow of the Corncrake) in Wicklow. He further stated that in Connaught and Ulster it is often made *tradhlach*, and his examples were Carrowntreila (Mayo) and Carrowntryla (Galway and Roscommon), the quarter-land of the Corncrake.

But the Corncrake's pedigree obviously predates literature. It is easy to imagine the first people arriving in skin boats on a fine spring day being hailed by Corncrakes calling from the marshy hollows of the coastal machair. It would have been sought after as food by these hunter/gatherers and, given its reluctance to fly, would have been easily caught in nets. It would have been local in its distribution, however, for the habitat which it requires was not widely available to it then: lowland Ireland was mainly a mosaic of dense forest and wetland. A.W. Stelfox found Corncrake bones in Keshcorran cave, Co. Sligo which, although not dated, were thought, from their location in the stalagmite floor layer, to be from post-glacial times – perhaps the remains of a Mesolithic meal.

From about 6,000 years ago the Corncrake would have occupied the new habitat made available by the *landnam* tree clearances and the resulting shifting agriculture of the first farmers.

Those stalwarts, who mattock-tilled the small fields and threw up stones to make the first walls, would have cursed the monotony of the 'crekking' from the plantain. It is a strange paradox that the bird which is nowadays being hastened to its demise by agricultural innovation was facilitated to spread originally by the same phenomenon.

Corncrake bones have been found at the 5,000-year-old habitation site at Lough Gur (Co. Limerick) and at Neolithic kitchen-midden sites on the Antrim coast, suggesting continuing use as a menu item, possibly after the advent of domestic poultry.

Archaeological evidence from Polynesia has revealed that an astonishing number (hundreds of species) of flightless rails – close relatives of the Corncrake – were hunted to extinction for food by the Lapita people on their eastward migrations across the Pacific, during our Bronze Age. These were mainly flightless species which, having abandoned the necessity to escape from their island idylls, had rendered themselves defenceless to human invasion. However, flightless or not, rails and crakes were easy quarry for the hungry hunter.

The Brehon 'Laws of Neighbourhood' (*Bretha Comaithchesa*), which are thought to be at least 1,200 years old, indicated the use of nets for catching birds (both for food and for feathers which presumably would have had multiple uses). Though they specified the circumstances including the lands on which they could be used and listed penalties for non-compliance, they did not specify the birds to be caught. There can be little doubt all the same that ground birds such as Corncrakes were high on the priority list.

Philip O'Sullivan Beare in his *Zoilomastix* gives us a rare picture of the Corncrake in Gaelic Ireland: '[The Corncrake] … which is called the little old woman, of the same size as the partridge; it accompanies the quails. It appears in Ireland through to the end of the summer and the beginning of autumn. It has a raucous call which it interrupts at the appearance of human beings. It makes its nest in the corn or nettles or hedges.' Not knowing the Spanish name (the *Zoilomastix* was written in Spain) he wrote in the margin: 'So the description of the traona [Corncrake]. Find out its name and do not delete.' With the abandonment of the Gaelic system in the 17th century and its replacement by the English statutes, laws relating to the seasons in which game could be hunted were enacted under a succession of monarchs. John Dryden, English poet of the later 17th century, referred to the Corncrake's culinary esteem: 'The rayle [*sic*] which seldom comes but on rich men's spits.' A game law of George III (1787) listed a number of bird species including the 'landrail' (the name Corncrake was first written in the mid 15th century) which could be legally hunted in an open season. Watters (1853) writing about its culinary value stated, 'At many of the dinner courses in the olden time, "reys" [rails] held a conspicuous place, – being valued at two pence each, ... which must cause us considerable surprise when we recollect the mallard was similarly valued.'

Nor was it seen only as a culinary item. In the days before medicine cabinets and pharmaceuticals, when ailing people looked to herb extracts and potions derived from the anatomy of animals and birds for remedies, the Corncrake came under scrutiny. The Reverend J. K'Eogh in his *Zoologica Medicalis Hiberniae* (1734) mentioned that anatomical parts could be used with similar efficacy to those of the Quail: 'The fat takes spots and films off the eyes, it is reported that the brain and dung cure the epilepsy: the flesh frequently eaten is a restorative in consumptions.'

The Corncrake was one of the birds to catch the imagination of English writers about Ireland and its curiosities. Giraldus, chronicler to Henry II on his visit to Ireland in the late 12th century, makes a succinct and revealing allusion: 'Hoarse and noisy *ratulae* are innumerable.' His *ratulae* are obviously Corncrakes; they may be readily identified with the family of ground birds, the ratites; and his laconic description, emphasising the call, leaves us in no doubt.

Tudor interest in Ireland focused on the country's natural resources, including her game. The frequent references to the 'rail' emphasised the abundance of the bird in relation to England. The Elizabethan chronicler Fynes Moryson, for instance, stated that Ireland 'abound[s] much more than England with rails'. F.J. Willughby in his *Ornithologica* (1676) gives us yet another (Anglo-Saxon) name: 'The rail or *daker-hen* ... this is very common in Ireland but much more rare with us.' Gwithers in his list of *Birds common in Ireland rare in England* endorses this consensus.

One 17th-century recorder of all things Irish, William Molyneux, had recourse to folklore in trying to explain the disappearance of the Corncrake in winter: 'The Irish report that rails towards winter turn to waterhens and in the spring the waterhens turn to rails. I did for several years endeavour to inform myself in this: and at last my falconer a little after Michaelmas brought me a bird that had exactly the body of a waterhen and the wings of a raile and I thought it very plaine that the raile had moulted the feathers of his body which came a darker colour; but had not moulted the feathers of his wings which confirmed me in the man's opinion. Especially considering a raile is a bird of a short and slow flight, and cannot make his passage out of the Kingdom as falcons, woodcocks etc. tho, yet they are never found in the winter among us.' This piece is interesting not only because of its colourful imagery in dealing with the mystery of migration but also because it contains echoes of the 'changeling' – so strong in human as well as birdlore in Ireland; the children of Lir/swan change and the barnacle/goose change are famous examples. In the Shetland isles Armstrong states that the sighting – as distinct from the hearing – of a Corncrake was thought to presage disaster, so unusual was it. In general, the folklore surrounding the Corncrake relates to it as a *sound* (as with the Cuckoo) rather than as a *bird*.

As regards overwintering in Ireland (a capability long attributed to Corncrakes) fact and fable are often interwoven. The 'Seven Sleepers', which includes the Corncrake, resonates with Celtic conviction; this folklist, 'The bee, the bat, the butterfly, the Corncrake, the Cuckoo and the Quail' is still passed on today in Tipperary (Gaynor, *per* Fox, 1999). It was undoubtedly originally in Irish and clearly predates scientific revelation.

Even the scrupulously reliable William Thompson included one or two unlikely secondhand accounts of overwintering Corncrakes in his *Natural history of Ireland* (1850). He also refers to one which was kept in captivity at Carrickfergus, summer and winter, and lived more than six years. However, there are a series of indisputable winter records in the *Irish Naturalists Journal*: one caught in a hedge, at Sydenham, Co. Down, on 8 January 1901, and others in Donegal and Galway (winter 1903). With hibernation the case is

61 *Corncrake and chicks at gate*

far from clear. Accounts of corncrakes being uncovered in compost heaps and suchlike can be found in 17th-, 18th-, and even 19th-century references. Some refer to the bird, on being taken into warmth coming miraculously back to life again, like a bat or a newt. Stanley (1857) recorded an incident like this from Monaghan. In the course of the removal of a large heap of manure which had laid undisturbed for a long time the workers found three Corncrakes:

> The birds on examination were found to be in a torpid state and were placed near a fire in a warm room. In the course of a short time a tremulous motion was observed in one of their legs and soon after a similar motion was noticed in the legs and wings of the whole which at length extended itself to their whole bodies and finally the birds were enabled to run and fly about the room. [Ussher and Warren (1900) made a bold attempt to clear up the confusion by suggesting that winter records, even into February, were simply 'hangers-on' and that there was nothing to show that they remained until March] … These winter birds have been found hiding in holes and chinks, but the torpid state attributed to them may have been that

simulation of death which the Corn-Crake is said to adopt when seized.

Ussher and Warren also provide us with vivid accounts of the migratory habits of Corncrakes. They mention frequent and abundant occurrences at lighthouses and lightships. They cited the fact that they were often killed striking lights as evidence which 'must satisfy even the most sceptical that this species can fly at high level and with great power and velocity'. They mention that no less than fifty Corncrakes were seen on the sea off south-eastern Ireland during foggy conditions in May 1867. Many later writers refer to the vulnerability of this bird on migration and to the fact that it is often found dead beneath electricity cables or other man made obstructions to its nocturnal migrations.

Nowadays it is impossible to comprehend how abundant the Corncrake was in Ireland – at least up to the time of Ussher and Warren. An account in the *Zoologist* highlights this and emphasises the destruction wrought by the advent of the mechanical mowing machine:

> Near Waterford these birds are so abundant this summer [1890] that during the cutting of a meadow of perhaps 4 acres a hundred or more were driven out, the last perch or so that was left uncut in the centre of the field being literally alive with them. It is a curious sight watching the crakes rising in two's and three's from the ever diminishing patch of grass or watch them scurrying mouse-like among the swathes of fresh-cut hay, while helpless young were destroyed in dozens by the machine. [A. Ellison]

The decline of the Corncrake has been well documented. It was obvious both to ornithologist and layman, so familiar a part of country life had it become. The disappearance from Britain (particularly eastern England) was underway by the latter half of the 19th century. In a number of European countries the decline was noticed in the early 20th century. Murton (1971) wrote, 'Once generally distributed throughout Britain and Ireland, the bird

[Corncrake] disappeared from East Anglia by 1900: from southern England by 1914; from Wales, England and most of Scotland by 1939.' In Ireland it was not so clearly defined. William Thompson noted a marked decline in the north of Ireland in the early 1840s, but by the end of the decade it had increased again. A contemporary, William Jardine, wrote of the situation in Scotland: 'In some parts it has decreased and without apparent cause: in the vale of Arran in south Scotland ten years since the bird was extremely common, its note being heard in almost every alternate field; at the present time it may almost be accounted rare, during last summer [1841] only one or two pairs being heard within a stretch of several miles'. Ussher and Warren on the other hand, were oblivious to an Irish decline in 1900. They stated that Corncrake numbers varied considerably from year to year and from locality to locality.

Kennedy, Ruttledge and Scroope (*Birds of Ireland*, 1953) recorded a general decrease throughout the country just previous to 1920 and noticeably by 1939. The authors' province by province, county by county account suggests a steady disappearance from the south

and east with less change in Connaught and west Ulster. Other Irish ornithologists recorded the decline after 1950. The joint IWC/BTO *Atlas of breeding birds* (1976) based on surveys carried out in the 60s and 70s graphically and accurately documented the status. Apart from scattered records from north-west England and western Scotland the Corncrake was more or less gone from Britain by then. In Ireland, though still present in every county, its distribution showed a decidedly north-western bias. Hutchinson's *Birds in Ireland* brought the story up to 1989 showing that by then its status was markedly fragmented with concentrations in parts of Fermanagh and less disturbed parts of the midlands and west.

Now, at the end of the century and millennium we may well be witnessing the tragic *finale*. The remaining stragglers hold on in their stronghold on the Shannon callows and along the country's north-western fringe. Even here where an extensive tract of habitat favoured by Corncrakes remains (hay meadows largely unaffected by artificial fertiliser, mown only occasionally and late in the season), it is difficult to see how such a small population (less than 200 pairs) can continue to be viable. In recent years great efforts

62 *(a) Corncrake distribution,* c.1970

62 *(b) Corncrake distribution,* c.2000

have been made by BirdWatch Ireland to conserve this seasonal meadowland and the associated traditional farming practices. It has not (to date) created a Corncrake sanctuary from which the beleaguered bird can regroup and repopulate. We can but hope that it eventually will.

Most observers point to the spread of mechanised mowing, first by horse, then by machine proper, over the past century as creating this unfortunate state of affairs (*Irish Birds*, 1994). Scything, even by efficient teams of men was a slower, less-traumatic process for the Corncrake and its offspring, which could usually escape to the untouched headlands at the meadow's edge. As the Ellison extract makes clear, the advent of mechanised mowing combined with an 'outside to centre' mowing regime at nesting time had a disastrous effect on breeding Corncrakes. The change from hay to silage, which since the early 1990s has been more or less ubiquitous, is seen as the most recent agricultural change to pressurise the beleaguered crake. Silage is much less attractive to wildlife in general due to its barrenness (a monoculture of perennial ryegrass boosted by artificial fertiliser and protected by pesticides and herbicides). There is no denying its quality as a fodder, however, and the bonus that it may be harvested two or three times, to the single cut of hay. Thus the cattle farmer is virtually guaranteed home-grown winter feed: previously, with hay there was always anxiety brought about by the vicissitudes of the Irish summer.

The conservation campaign has centred around encouraging (and compensating) farmers for putting off the cutting of those remaining hay-meadows till July at the earliest and using a 'centre-out' method to allow the Corncrakes and their young to escape rather than be corralled and killed in the middle of the field during operations. The scheme has met with a measure of success both in saving Corncrakes (the population of about sixty calling males in the callows has remained fairly stable for a number of years) and in introducing farmers to conservation methods. But while there will undoubtedly be long-term benefits to the ecosystem and to its sensitive species as a result of this scheme it does not seem to be reversing the overall declining trend in Corncrakes.

So, what is responsible for this apparently inexorable decline? Is the Corncrake, as we are led to believe, simply a casualty of modern agricultural practice – one more example of bio-diversity depletion caused by progress and the insensitivity of the Common Agricultural Policy? The careful chronology of the decline, occurring as it did in parallel with agricultural mechanisation and modernisation, would seem to leave us in no doubt. However, as in most environmental questions, it may not be as simple as that: a wider look at the circumstances suggest that it could be multi-factorial.

Michael O'Meara's comprehensive inquiry (*Irish Birds*, 1979) pointed out that no correlation could be found between silage-making and the decline and that birds were now absent from areas where there was no such activity and little mechanisation. Certainly this holds for the Burren in North Clare, which did not see the advance of silage production until the mid 1980s yet Corncrakes were vacating the ideal habitat of the Burren's hay meadows twenty years earlier. Though O'Meara concluded that the reasons for the decline 'remain obscure', he did point to contributory factors outside Ireland altogether. Hazards such as those encountered on migration in the wintering quarters were mentioned. Significantly (and this has been highlighted elsewhere) the Sahel region, the 'shore' of the southern Sahara, has been coming increasingly arid and unwelcoming to migrant birds since the early 1960s due to socio-environmental reasons. Crashes in the populations of other migratory birds – notably the Sand Martin and the Whitethroat – have highlighted the hazards of increased desertification. Could it be that the distance which Corncrakes now have to fly to find sanctuary in Africa after their prodigious migration has stretched its population sustainability to breaking point? Add to this the widespread trapping of Corncrakes, carried out the Mediterranean coast of north Africa – thousands (14,000 in 1994, RSPB) are netted for food every autumn in Egypt – and the proneness of the bird to accidental calamity (power-cables, lighthouses etc.), a picture of some complexity emerges. Stephen Mills (*Nature in its place*, 1987) specifies availability of food as being a major factor: 'Corncrakes eat everything the meadows have to offer – crickets, spiders, beetles, caterpillars, flies, worms and seeds. The new farming systems reduce the variety of most of these. They must use a large amount of energy on their African journey. If they, and especially their inexperienced young birds, set out on empty stomachs they will never get there.'

Increasing wetness and other climatic fluctuations (as a consequence of global warming?) have been suggested as contributory factors in the decline, though they tend to be relegated to secondary considerations. How much of this is due to the fact that they cannot be easily quantified, or their refusal to fit neatly as parameters for statistical analysis? Is there a danger that, in the in the desire to solve the Corncrake problem, we deny ourselves a view of the wider picture by focusing on those aspects of the bird's lifestyle that we can monitor? The Royal Society for the Protection of Birds have made brave efforts to enlarge upon our knowledge of the Corncrake's migration patterns and its African wintering grounds (Stowe & Becker, 1992). However, the detail which they assimilated about the winter range – mainly a handful of south and south-eastern African countries – greatly outweighed that regarding their migrations within the continent. Some valuable data, from Morocco and southern Algeria, when taken in conjunction with a number of ringing recoveries (from France and Spain) suggest a south, south-east flyline from Britain and Ireland. Stowe & Becker are sanguine about the wintering quarters, in southern and south-eastern Africa stating that 'no conservation action is warranted'. However, their migration summary ('Less is known of the problems for migrating birds but these could be severe') hints at factors which may be increasingly impacting and moving beyond the realm of our control to slow down or reverse. The importance of such extraneous influences may unfortunately not become evident for decades or even centuries to come – too late for the Corncrake.

Could it be that the much depleted western population of the Corncrake is experiencing isolationism in its African winter quarters, as it is now in its summer quarters in Ireland? Perhaps the remnant Irish (and Scottish) population overwintering separate from the bulk population are therefore more vulnerable to the pressures that adversely affect isolated populations. Ringing might

have clarified this and other questions, but recoveries have been so few and scattered that the issue seems destined to remain obscure.

The Tassili-n-Ajer region of south-east Algeria which lies on the western migration route of the Corncrake holds a remarkable cultural record of change, in the wonderful prehistoric rock art to be found there. The bulk of the paintings dating from 12,000– 8,500 years ago are of animals (Giraffes, various antelopes, Hippos, Catfish), savannagh and wetland creatures long gone from this now-barren wasteland. Pictures of chariots from c.3,200 years ago show that the landscape was still pre-desert then. The first camel drawings clear indicators of desertification date from c.2000 years ago. The ramifications are compelling: this part of north Africa changed gradually from fertile grassland and wetland to parched desert, mainly as a consequence of natural climatic change, though probably accelerated by human activity. Such dramatic, albeit long-term, change must have impacted on the western population of the Corncrake either by limiting the food supply available in migratory landfall circumstances or by pushing them south to more acceptable habitat. The eastern population (by far the majority nowadays) has probably always had a shorter, less-hazardous route from the steppes to suitable savannah in east Africa.

In the broader consideration, what is the evolutionary lifetime of a species like the Corncrake? It is thought that in the normal course of events the average lifespan for a species of bird is of the order of two million years. This assessment is based loosely on an analysis of the fossil record which comprises many species (like the Moa of New Zealand) which are no longer around. While this theory is unavoidably speculative, there have been corroborative examples in, for instance, the Dodo of Mauritius. It might be argued that the Dodo did not have to become extinct: it was driven to extinction by exploitative man. But this could also be seen as an example of a species which having arrived (as a wind-blown pigeon?) on a remote island devoid of natural predators proceeded to evolve into a highly specialised, flightless form which was doomed to collapse one way or another. The California condor, presently one of the rarest birds in the world and on which millions of dollars are being spent to keep it from extinction, appears to be dying out quite naturally. Its progenitors' fossils have been found in the La Brea tarpits, suggesting that condors were in their heyday when the continent was full of a wide range of large mammals which left abundant carrion in the landscape. Nowadays carcasses are being left out specially for the condors to enable them to continue to live in the style to which they have become accustomed. The huge expenditure on the condor has been justified on sentimental rather that ecological grounds. The western States cannot afford to let such a powerful symbol of the 'wild west' simply slip into oblivion without a fight.

Whatever about the *rationale* behind the conservation of a relict sedentary bird, how do you begin to protect a bird that migrates 6000 miles each year? The plight of the Corncrake, while hardly warranting the same urgency as the condor, is nevertheless acute. At the present rate of decline it could be extinct in a century.

In 1987 the National school at Creagh, near Ballinasloe, Galway, won the ESSO schools wildlife competition. Their project 'Bring back Coreen's Corncrakes' investigated the disappearance of the bird locally. It surveyed farmers in Coreen, mapped where Corncrakes used to nest, researched their lifestyle requirements including breeding and identified actions which could help save Ireland's remaining Corncrakes. Significantly, there was widespread emotional response from the landowners, who rued the disappearance of what was for them more than just a bird. It is fitting that such a project should have emanated from the edge of their traditional stronghold in the Shannon basin.

One notable outcome was the high level of cooperation between the young conservationists and the local farming community who, in the inclusive atmosphere of the project, responded positively. The Corncrake grant scheme which, through monetary incentives, seeks the cooperation of farmers on whose land the birds nest, has met with similar cooperation (*Irish Birds*, 1998). The Corncrake has thus become a catalyst for cooperation between perspectives which have traditionally been more often confrontational.

The Corncrake will never return to its former ubiquity: the landscape has changed too radically, too permanently for that. The finest gesture we can now afford this venerable and quintessential Irish bird, in Ireland, is to conserve its callow stronghold for the wealth of wildlife which it supports and in so doing let nature take its course.

63 *Corncrake and chicks*

BIRDS GAINED

Our bird tally of more than 400 bird species seems high for a smallish island like Ireland. This figure is not representative, however, since it comprises about two thirds scarce migrants or rare visitors and is more a comment on the success of bird seekers than on the birds. Our 130 or so 'familiar' birds are a better gauge. These are not only the all-year-round residents, like crows and sparrows, but also birds that come here to nest (Swallows and Cuckoos), some which spend the winter here (northern thrushes, waterfowl) and a few well-known birds of passage.

On the above basis we have a small enough complement of bird species, about two thirds of the British total but only about a third of that of western Europe. Despite this impoverishment, Ireland is strong in certain birds. We are probably as well endowed with seabirds and wintering waterfowl as anywhere on a similar latitude. The reason for this is largely geographical – our proximity to the ocean and to the western European migration flyway. But it also has to do with the availability of good quality habitat. Ireland's relatively undeveloped rocky coastline is attractive to many coastal species which come there in their thousands to nest. Our coastal estuaries, still largely free from the detrimental effects of industry and inland wetlands, remain attractive to myriad wildfowl and wading birds.

McArthur and Wilson (1967), recognising a relationship between the number of bird species on an island and its size, came up with some interesting theoretical insights. They found that the island bird population tends to be constant – immigration of new species balancing extinctions, over time. They proposed that species with low populations (habitat-specific species, for instance) would tend to die out due to competition from invaders establishing footholds and that the actual rate of invasion by new species is dependent on the distance from the *source* population rather than the *nearest* population. Most significantly, the rate of extinction of species is proportional to the size of the island: it is higher on small, isolated islands.

Hutchinson (1989) viewed the Irish situation in 'Why are there fewer species in Ireland?': 'Firstly it appears that a number of summer migrants [Pied flycatcher, certain warblers etc.] have not colonised Ireland because they are at a competitive disadvantage from year-round residents [Goldcrests, tits, etc.].' O'Connor (1986) has shown that only 16 of the 40 migrant species (35%) regularly breeding in Britain also breed in Ireland, but 86 of the 125 resident species (69%) do so. He also wrote that the absence from Ireland of certain sedentary species such as Tawny owl, woodpeckers, Nuthatches, Marsh and Willow Tits was likely to be due to their inability to reach Ireland at all – a vindication of McArthur & Wilson's 'source/distance' theory. As Hutchinson also noted, this was later challenged by the biogeographer David Lack (1969) whose belief was that the Irish Sea did not represent a barrier to these birds and that other factors were at play.

Hutchinson's summary indicates that the issue as presently understood is far from clear: 'The reasons why small islands in general have fewer species than large islands are still debated among scientists and Ireland has been used as an example by several. The current thinking indicates that there is not just one simple explanation but that several factors are involved …' Two of these factors are habitat range and climate. These present reasons why direct comparison with Britain is difficult to substantiate. Ireland has a good variety of habitats, from wetlands to mountains, from woodlands to bog … virtually all under the influence of an Atlantic climate. The north of Scotland and the south of England have habitats other than those found here – pine forest and mountain tundra, dry heath and chalk downland, for instance. These extremities are also influenced by climatic conditions other than the Atlantic. This has meant that certain birds, Ptarmigan, Dotterel, Crested Tit, skuas, etc. – birds with strong 'northern' affinities – and Red-backed Shrike, Dartford Warbler, Wryneck – 'southern birds' – nest in Britain at the southern and northern extremities of their range. With very few exceptions (Red-throated Diver, Nightjar?) Ireland lacks similar examples.

The birds which, in modern times, have come to Ireland and established themselves (perhaps at the expense of, or having taken advantage of, niches vacated by others) are well known. It is likely that, with the disappearance of those which comprise the case studies of this book, the process of invasion and niche establishment was made easier. The widespread habitat modification that enabled this to happen is a relatively recent phenomenon, most having

65 *Stock Doves*

occurred since the 17th century. It is likely therefore that the invaders which have made a success of their colonisation since that period represent the majority of those which have come, overall.

The Magpie is probably the first of these. Giraldus's 12th-century comment that 'There are no magpies [in Ireland]' was corroborated by other writers including Stanihurst, in Elizabethan times and Moryson as late as 1617. There can be little doubt that they were correct, for the first record, in the barony of Forth and Bargy, Wexford (*c*.1676), left an impression locally (Leigh, 1684). A 'parcel of magpies' appeared after a storm and as unfamiliar blow-ins were undoubtedly soon the subject of gossip and folklore. The Magpie was well enough established by 1684 to be listed under 'Birds rare in Ireland and common in England' by the Dublin Philosophical Society. It was well known to the 18th-century writers, Smith (Cork) and Rutty (Dublin). Smith mentioned that by the mid 18th century it was a common Irish bird.

The Mistle Thrush was apparently unknown in Ireland until 1808 when one was shot in Co. Antrim. It was recorded in Co. Tyrone in 1820. Breeding was confirmed by 1836. It has since become a common and widespread resident throughout the country. The Stock Dove is another newcomer. Unknown until

64 *Magpie*

66 *Head of Fulmar*

1875 it was breeding in 1877 and now, though it remains thinly distributed, it is fairly widespread in Ireland.

Two coastal birds, with northern distributions (the Fulmar and the Eider Duck) are newcomers – the Fulmar to Mayo in 1911; the Eider to Donegal in 1912. Both are strictly coastal species. The Fulmar, which arrived on our shores behind returning fishing boats, has been more successful as a coloniser and is found nowadays all around our coastline. The Eider's spread has been slower. While it is well established along Ulster's coast, it shows no signs of following the Fulmar around our coastline.

The glamorous Goosander is also a newcomer to Ireland. Over the closing decades of the 20th century it has bred regularly in Donegal and latterly in Co. Wicklow. This reflects a trend which, since the mid 19th century, has seen the Goosander spread south to colonise Scotland and extend to northern England and Wales. Along with the Red-throated Diver (also a recent rare coloniser in Donegal) and one or two wading birds, which have nested as one-offs or sporadically, it represents a suite of subarctic species which have managed to gain a tenuous foothold here. The Goosander shows signs of being more successful than most.

Our most dramatic invader and coloniser – the Collared Dove – came, not from the north, but from the deep south. The story of its 'explosion' out of its homeland in Asia Minor north-west across

Europe, is well known. Within a few years it had colonised much of western Europe. Having arrived in Ireland first in 1959 it made itself at home in Leinster and Ulster and began to breed almost immediately. Nowadays it is a firmly established resident, occupying villages and farmsteads even in the extreme west of the country.

Other birds with southern distributions have also successfully established colonies in Ireland. Since the 1980s Reed Warblers, formerly scarce passage migrants, have begun to breed in suitable reedbed habitats along the south coast. They have not only prevailed but have consolidated their position and continued to spread. It is conceivable that they will, in a decade or so, occupy suitable habitat throughout Ireland. Lest we talk too soon, however, let us consider the Bearded Tit, which also came to breed in a few reedbeds in the south and east and has since more or less died out. Cold and wet winters may have put paid to their chances. The most recent coloniser is that beautiful white heron, the Little Egret. It was known here only as a rare visitor until, in 1997, a flock arrived and established a colony in Cork. Since then it has bred every year and has spread locally.

67 *Collared Doves*

There are a few other less obvious examples of birds that were formerly scarce winter visitors or passage migrants which have become residents here. These include the Siskin, the Crossbill and the Tree Sparrow (all since the early to mid 19th century). The Irish model would thus seem to substantiate at least one of island biogeography's theoretical assertions – that island populations tend to reach equilibrium over time, with invader species tending to replace extinctions. Over the past four centuries we can find almost as many colonisers to replace those that have gone.

We must not forget a few species of birds which have by way of human intervention become residents here. The best known is undoubtedly the Mute Swan. It is hard to believe that the familiar orange-beaked swan of our canals, rivers and lakes was probably unknown here prior to Anglo-Norman times. The swans of the Children of Lir, the placenames associated with swans (Ballyalla, Co. Clare) and the masses of swan bones uncovered at crannog sites are invariably those of wild swans – mainly Whooper Swans, winter visitors from Iceland. Interestingly, the Whooper Swan has, over the past few years, stayed to nest here in one or two localities. The Mute

Swan was probably initially brought into Ireland as food, but in the more settled, Jacobean period it would have been a mandatory decorative addition to the artificial ponds of the demesnes.

The Pheasant was also introduced into Ireland in the latter Middle Ages. A native of Asia, it was well known and savoured by the Normans but may not have been released in Ireland until Tudor times. It was certainly well known by the early 17th century in some Ulster estates. Nowadays its raucous call is a familiar sound in most parts of the country, and it has remained an established member of our avifauna. It is debatable, however, whether the Pheasant would continue to survive in the wild here without the ongoing release programmes of the gun clubs.

Canada Geese have been introduced to a number of wetlands both in the north and south of Ireland. Those in Strangford Lough are feral in that, having been originally introduced and released, they have reverted to the wild state, sustaining their numbers and reproducing without human intervention. Hundreds are now found on some of the Fermanagh lakes, where they were introduced as a wildfowl collection decades ago. Nowadays small flocks are found as peripherally as Waterford and Limerick, and it seems certain that, if measures are not taken to control their spread, these robust Nearctic geese will eventually become as familiar as Mute Swans in many places.

The Ruddy Duck, also a Nearctic native, was introduced into wildfowl collections in Britain from where it escaped. In recent decades it has established itself as a regular breeding bird in the wild, particularly in northern Ireland. The majority are found in the Lough Neagh region. The Irish breeding population has been estimated at around 50 pairs (Wells et al., 1998). Ruddy Ducks have turned up in many locations throughout the island but have not become established to the extent envisaged on initial breeding in the north.

While our birdlife may be described as impoverished in terms of numbers of species, actual bird populations are often quite high. Our cities and towns are as full of Starlings, pigeons and sparrows as their European counterparts. Our dumps and docks are just as alive with gulls. Our golf-courses and recreational areas support abundant common 'ground birds' like pipits and wagtails. Forestry

68 *Little Egret*

plantations, though especially weak in variety, hold large populations of Goldcrests and tits. Irish gardens, whether suburban or rural, have a consistent suite of about a dozen residents and are secure sanctuaries some of our most familiar birds, like Robins, Chaffinches and Blackbirds.

Most Irish people see birds peripherally, in the course of unrelated activity – going and coming from work or in recreation of some form or other. Thus crows, thrushes, tits, finches, gulls, pigeons and a few other families are the ones most frequently encountered. Some water birds – Grey Herons, Moorhens, Mallard, Grey Wagtails – also catch the eye from time to time. All in all, however, the majority of people can identify with as few as a score of familiar birds. Any of the more than a hundred others on the list, when noticed, are regarded as 'unusual' despite paradoxically being also 'familiar'. This, of course, has more to do with the degree of awareness of the observer than the unobtrusiveness of the bird.

Things are changing, however. An increasing number of people are noticing increasingly more birds. There are nowadays more people aware of and watching birds in Ireland, north and south, than ever before. BirdWatch Ireland (formerly the IWC) has a steadily increasing membership (presently approaching 8000). Many of these are 'twitchers' – birdwatchers preoccupied more with the unusual than the commonplace. But, having participated myself, I can testify to the appeal of 'twitching' – a kind of benign hunting, spiced with the thrill of discovery. The activity has the beneficial effect of honing the senses and increasing awareness.

BirdWatch Ireland is also the largest nature-orientated organisation in the country, echoing the massive popularity of bird-watching in Britain (where the membership of the Royal Society for the Protection of Birds is of the order of hundreds of thousands). While this burgeoning of interest is marvellous to behold and obviously reflects the intrinsic attractiveness of birds, it is not so clearly seen in other areas of the natural world (the Slug Society is not doing so well, I hear). It is hardly ideal to have so much nature enthusiasm channelled through a particular conduit. The obvious danger is a selective approach to conservation – birds over beasts? A more inclusive forum would have a stronger philosophical basis and would result, in the longer term, in a louder, broader voice for conservation. There is, nonetheless, a resolute ongoing drive by the bird societies to conserve habitats – primarily for birds – but also for their overall ecology. BirdWatch Ireland has branches throughout Ireland and a network of reserves and sanctuaries, many with observational hides. These facilities have greatly increased public awareness of our birdlife. Many birdwatchers are committed 'garden birders'. Despite the obvious restrictions of scope some pioneering ornithological work has been done in this humble context; also, through the undemanding everyday recording of birds frequenting gardens observers contribute to a vast information bank of census data.

This may enable us in the future to find out which, if any, of our familiar birds are (like the Song Thrush in Britain) becoming less familiar.

CONSERVATION

In most modern cultures, birds and their welfare are invariably low on the legislative priority list.

Conserving birds for reasons other than maintaining stocks for hunting and therefore for food is an historically recent phenomenon – more recent in Ireland than in many other European countries.

Birds were important enough in early Gaelic society to be mentioned frequently in the Brehon laws. The laws of neighbourhood (*Bretha Comaithchesa*), which concerned the obligations of neighbouring farmers, have many allusions to birds, often relating to offences caused by pets and poultry. *Bretha Forma* included references to wild birds caught for food, legal ownership of the flesh and feathers and penalties associated with non-compliance (Kelly, 1997). Exempt from penalty were small birds, presumably caught for caging as songbirds.

The triads (early verse) and other literary snippets which contain bird references sometimes present birds in an anthropomorphic manner. This may be benign or humorous (Corncrake) but can also be sinister (Raven); in general they reflect the spontaneity and wit of the observer. Kelly makes it clear that early Irish law regarded many predatory birds as undesirables. O'Davoren's glossary states that this is because of the depredations of young pigs and poultry, suggesting larger rather than smaller predators. On the other hand, eagles and other birds of prey were commonly kept as pets – though not for falconry, which was not apparently practised by the Gaelic aristocracy.

The early Irish nature poetry of the ascetic monastic tradition is liberally sprinkled with bird references. Song birds such as the Blackbird and the Cuckoo feature highly. The sagas too indicate a close association with certain birds (wild swans in the 'Children of Lir'). The impression conveyed in many cases is of quiet appreciation, if not admiration of the birds in question.

When the Norman king Henry II came to Dublin Castle in the 12th century he feasted on Irish Cranes. He was, according to Giraldus (*Expugnatio Hiberniae*), entertaining (or being entertained by) Irish chieftains who were appalled at the idea of eating Cranes. There is also a long-standing tradition of not killing swans for food (though their bones have been found at crannog middens), which must have considerably amused the gastronomically liberal Normans when they settled here. Reading between the lines of this early material it can be deduced that, while birds were largely viewed as a commodity, there was also a certain tolerance and even affection for many. Exploitation seems to have been limited to the circumstance of necessity rather than indulgence. This is not to say that anything approximating to a conservation ethic existed in Gaelic Ireland; rather, the insular, clan-based structure had neither the wherewithal nor the mercantile drive to embrace the exploitation we associate with the expansionism of the Early Modern period.

It is clear from the series of Irish statutes enacted from Dublin Castle that birds were protected only insofar as they were important as game. There are many references to the open seasons on birds such as Grouse, Woodcock, Partridge, Rails (Corncrakes) and Quails in the statutes of Edward, Anne and the Georges. It is hard to believe that these laws had any affect in limiting or controlling bags outside the walls of the demesnes. As regards the Cock of the Wood (Capercaillie) its eleventh-hour legal standing was of no avail in saving it from eradication.

Complementing the game lists were the vermin lists – mainly birds of prey and crows. The removal of the threat to game from would-be predators was as important a factor as the actual rearing and conservation of the stocks. The brief, non-specific lists in the game laws point to the manner in which control was essentially decentralised into the hands of landowner and gamekeeper. The 'vermin books' of the larger estates make woeful reading. Arthur Stringer's *Experienced huntsman* (1714) provides us with a marvellous insight into the ruthlessness of an efficiently run Ulster estate of the Georgian period.

The attitude of the burgeoning peasant tenantry must also have hardened. The familiar tolerance which might have accompanied predatory birds and scavengers degenerated to abhorrence. Certainly the impression which comes across from the diaries of Humphrey O'Sullivan (1827–35) is that anything which threatened interests or livelihood was regarded as an enemy. On this issue the privileged and underprivileged were united. Paradoxically it has been pointed out that eagles, particularly Golden Eagles, managed to hold on to their traditional corrie and crag eyries despite the approach of the 'lazy beds' of a burgeoning peasantry up the mountain sides. However, the eagles posed no threat to the livelihood of potato farmers, who probably adopted a 'live and let live' attitude to avian predators. O'Toole has pointed out that Golden Eagles continued to occupy territory in the Scottish Highlands in close proximity to (often overlooking) the summer shielings of the upland grazers.

In the social turmoil which afflicted Ireland throughout the 19th century there are few enough allusions to birds other than those of their recorders – Templeton, Thompson, Watters etc. – who were facilitated in their work by virtue of their social class.

The catching of songbirds was part of the fabric of rural society in Ireland. Though nowadays condemned as cruel, it had in its day a certain bucolic innocence. The cheerful singing may have held deeper meaning such as defiance in the face of adversity. It connected with other traditional bird-motifs – the Wren as the king of the birds (the elevation of the underdog); the salvational motif of the Robin, its breast red with the dripping blood from the

WHEREAS the Game hath been much deſtroy'd in the Eſtate of the Right Hon. the Earl of Antrim in the County of Antrim: This is to give Notice, that the Law will be ſtrictly put in Force againſt any Perſon (being not duly qualified by Law) who ſhall come into any of his Lordſhip's Manors, either with Gun, Net, or Dog, to kill or deſtroy either Partridge, Hare, Grouſe, Quail, Pheaſant, or any ſort of Game, or Woodcock or Snipe: It is alſo expected, that there will not any Perſon (tho' qualify'd) Sport on any part of ſaid Eſtate, without proper Leave and Authority.
Dated this 16th of October 1749.
By Order of his Lordſhip,
WILLIAM HARRISON.

69 *Game proclamation, 1749*

cross. Caged birds thus represented significant comfort symbols to countryfolk, to the chagrin of the early ornithologists, who saw the practice in purely practical terms, as a threat to the survival of some species. Templeton (early 19th-century) was concerned that the trapping of Woodlarks was causing serious decline in their numbers in the vicinity of Belfast. A turn of the century photograph of Kinvara shows the little cages outside the doors of many of the houses, an indication of how common the practice remained in rural areas. My own granduncle, an ex-RIC man from Dunmore in Galway, was known as the 'birdman' due to the many cage-birds which he kept in his house, in the early days of the new state. He saw nothing illegal in it. An old man interviewed on a recent RTE programme recounted how, as a child, he caught and caged birds in the Ballyhoura hills. He described how the birds were trapped by the feet using an extract of linseed oil, a substitute for the more common Holly sap or lime. Undoubtedly there was an element of cruelty in the practice and the occasional bird surely perished, but it is unlikely that bird-trapping was responsible for the extermination of any of our native species. The favoured species were Linnets, Yellowhammers and Goldfinches, presumably for their double attraction of song and looks. While Goldfinches continue to be common (especially in the west) – wherever there is rough ground – both the Linnet and Yellowhammer are less common than formerly. This is nothing to do with bird-catching, however, for the practice has all but died out: it is more to do with the reclamation of commonages and increased use of chemicals on the land.

The Bird Protection Act of 1931, a geographical extension of the 1930 Northern Ireland Bird Protection Act, was the first piece of legislation to deal with the threats to our wild birds. The Act specifically outlawed the catching and caging of songbirds, egg-collecting and the use of gin and pole-traps to kill birds of prey. Also, specific shooting seasons were prescribed. The practices of trapping and egg-taking lingered on, however – the former simply because of the impossibility of ending suddenly such an endemic activity; the latter because of the lucrative black market which had developed among the leisured class. This eventually waned, in part due to its own destructiveness: the most sought after birds had been wiped out. The legislation had come too late to save the eagles, the Marsh Harrier and the Buzzard, all of which fell victim to greedy oologists.

Ironically it was the Victorian landowner stratum, the repository of most of the collectors, which produced our first conservationists. Oliver Grace, who owned extensive wetlands on his Co. Roscommon estate at Mantua, conserved the Bitterns (which bred there) by instructing visitors not to shoot at them. His methods may have been crude and his motives ulterior – Bitterns would have been quite an attraction for guests – but he was nevertheless acting as a conservationist (unsuccessfully as it turned out). The vast fenlands of Lord Castletown's Co. Laois estate held Ireland's last Marsh Harriers in the early part of the twentieth century. However self-interested it might have been (Castletown entertained visitors to shoots on the fen), his plea to the newly created council to have the wetland preserved was visionary. Unfortunately for Castletown and the Marsh Harriers, the zealous servants of the new state did not share the sentimental rantings of a member of the former, despised establishment, and the wetland (more than a thousand acres) was summarily drained.

The Royal Society for the Protection of Birds was formed in Britain in 1891 by individuals concerned about the near-extinction of herons and grebes for the feather-hungry millinery industry. The Irish Society emerged in 1904, though its formation may have been more to do with the flourishing of natural history as a 'respectable pastime' than as a conservation organisation *per se*.

Certain individuals, notably George Humphreys (despite being an avid egg-collector) and C.B. Moffat, did much for the cause of bird protection in its formative years. Moffat, prior to 1945, was particularly concerned with conserving the Red-necked Phalarope in Co. Mayo at its only Irish breeding site. The on-going protection afforded, with the appointment of a local warden, was a benchmark in the protection of Ireland's threatened birds (Hutchinson, 1989). Other sites such as the Bull Island, Dublin, and the Wexford Wildfowl Reserve followed. Protection for the former was helped by Father Kennedy's delightful book *An Irish sanctuary* (1953) which publicised its importance as a bird habitat. But despite these isolated successes bird conservation was slow to take off in the Republic. It lacked the ground swell of public opinion so well established in post-war Britain (and in the North of Ireland).

One factor, in particular, militated against this. The fact that the teaching of nature had been more or less dispensed with in the National school curriculum meant that the Republic's mid 20th-century generations grew up with little educationally endorsed empathy for birds. This point is well articulated in Foster's *Nature*

70 *Black-necked Grebes*

in Ireland (1997). A rural population stimulated by developmental zeal and a new Arterial Drainage Act (1945) set about converting the soggy catchments and residual wetlands into viable agricultural land. Bord na Móna echoed this enterprise in the exploitation of the bogs. Over the period of the 1960s and early '70s when most of the radical drainage was carried out, more than 100,000 hectares of wetland was affected. The adverse effect on vulnerable birds such as Greenland white-fronted Geese and Red Grouse was neither gauged nor considered. Few voices were raised to challenge the exploitation. In the early 1970s, An Foras Forbartha, in listing 'Areas of Scientific Interest', highlighted the threat. It was not, however until *Land drainage policy in Ireland* by Bruton and Convery emerged in 1982 that scaling down became effective. It took intervention from the Dutch government and the formation of the IPCC (Irish Peatland Conservation Council), at about this time, to save our last remaining raised bogs. Bord na Móna, so successful in its exploitation and employment brief, would have otherwise exhausted them by now.

It is unlikely that any of our vulnerable birds (with the possible exception of the Black-necked Grebe) actually died out as a direct consequence of the mid 20th-century drainage, but the habitats (fens and bogs) available to wetland birds in general were greatly reduced. Conservation was carried out locally but on a widespread basis by gunclubs. Many small wetlands and copses which might have otherwise disappeared under the bucket of the drag-line or the 'dozer were saved for and by these clubs. The control of game and predators which formerly had been in the hands of the big landowners was now distributed far and wide among the rural community. It is debatable whether greater protection was provided in the early stages. The estimated number of gamebirds shot rose from 189,370 in 1965 to 446,673 in 1980 (An Foras Forbartha, 1985). Certainly, native gamebirds such as Red Grouse and Grey Partridge continued to decline, though there may have been other, extraneous factors, involved. The coming together of the 28 Regional Game Councils in 1969 to form a national association (NARGC) created a powerful voice for sustainable hunting.

71 *Grey Partridge pair*

It is doubtful whether a significant climate for conservation would have existed in this country were it not for our membership of the European Community. It has been a two-sided relationship, of course. The widespread land modification (habitat destruction?) carried out under the grant-support system has been a boon to the farming community. Unfortunately (as in the Burren) considerable damage was caused to hitherto unmodified areas before a proper assessment could be made of their conservation potential. Stimulated by the European Conservation Year, designated by the Council of Europe in 1970, the inventory of some 2000 Areas of Scientific Interest (ASIs) was drawn up by (now defunct) An Foras Forbartha. (In fact the ASIs were found to have no legal standing when challenged in a famous court case.) The Forest and Wildlife Service became the body responsible and the Wildlife Advisory Council took on the task of championing the wildlife conservation cause. The liaison of forestry and wildlife has been subject to considerable criticism based on the radically different outlook of each of these elements – forestry, essentially an agrarian business; wildlife, a 'quality of life' enterprise. It has to be said, however, that despite the adverse environmental effects of alien conifer plantation, certain birds have benefited. Nightjars and Hen Harriers, birds listed as endangered or rare in the *Red Data Book* 2 (Whilde,

1993) have nested in Irish forestry plantations, though usually at sapling stage. They tend to vacate them when the trees form a canopy. The Wildlife Act, giving statutory power to the fragile conservation efforts which were under way, came into being in 1976. Eighteen of Ireland's native birds were listed as 'threatened' under the Act and protected by order (Cabot, 1980). While most of the others were also afforded protection a number of licences and permissions were granted which enabled individuals to 'capture/kill certain wild birds for educational, scientific or other purposes' and 'control protected (non-game) species causing damage'. This latter clause enabled birds such as Grey herons or Cormorants visiting fish-farms to be killed with impunity. Licences were also granted to take birds of prey for falconry or in exceptional circumstances for captive breeding. Some ornithologists (not without reason) objected to any tampering with our vulnerable wild stock of raptors which were already under considerable pressure due to chemical sprays and grain dressings which were contaminating their seed-bird prey.

Reservations aside, the Wildlife Act was, in its day, a revolutionary document. Having served us well, for more than twenty years, it has been found in recent years to be in need of revision. Hedgerow destruction and the continued use of non-discriminating poisons on the land are two of the issues which have brought this to light.

The Birds Directive (1979) invested all European member states with the responsibility to protect all wild birds and their habitats. Those bird species which were 'giving cause for concern' were specifically listed under Annex 1. They were grouped under the following:

ENDANGERED (birds in danger of extinction)
VULNERABLE (birds believed likely to become endangered)
RARE (birds with small populations, at risk, not listed above)

Two other designations, INDETERMINATE and INTERNATIONALLY IMPORTANT, are self-explanatory.

Under EXTINCT the criterion is, 'Species not definitely found in the wild over the past 50 years'. Only six species were listed for Ireland in 1979: this is likely to double over the next 50 years with birds such as Corn Bunting fading away. Twenty-eight others are listed as ENDANGERED, VULNERABLE or RARE. Again this will grow without significant conservation intervention. This will certainly involve not only the protection of the birds themselves but also the habitats in which they are found.

A network of protectorates specifically for birds (SPAs) has been designated under the Birds Directive. More than twenty of these have been identified in the Republic. These have been complimented by a series (under an arrangement of the Nature Conservation Committee of the UK) in Northern Ireland.

The responsibility for bird (and wildlife) protection has since been transferred to the Office of Public Works which has refined the designations further. It now lies with the National Parks and Wildlife Service – the NPWS. With the ASIs being converted to National Heritage Areas (NHAs), it was thought that statutory recognition would have been a formality. However, much to the disbelief of everyone in the conservation world this has not happened and the delay has resulted in more regrettable habitat damage. Legal complications have been cited as reasons for the delay but, whatever the problem, it must be resolved *post haste*. One of the unfortunate side-effects of the emergence of so many legislative instruments is that the area of nature conservation is now awash with acronyms. This has done nothing to assuage the misgivings of a bewildered rural community which is at last coming round to seeing the wisdom of the programme.

Armed with these and other European and international instruments listed below, Ireland's wild birds have at last been given a solid conservation mandate. While conservation is not out of the woods yet, and factors beyond internal control could disrupt the programme, adequate structures will soon be in place.

BirdWatch Ireland are prime movers behind an exciting new scheme (1999) which may aid the Corncrake in its Shannonside stronghold. Under LEADER II, the EU-backed rural development programme, a pilot scheme has been devised to find the best conservation strategy for the Shannon callows. Significantly there is a high level of local cooperation, with the Banagher Development Group involved from the outset. The enthusiasm for this pilot project augurs well for the future and may become the model for others throughout the country.

The revision of the 1976 Wildlife Act is eagerly awaited. Having been proposed as early as 1997 it remains (May 1999) at Bill stage despite intense lobbying by the major wildlife organisations. Most people are hopeful that it will appear before the commencement of the new millennium.

While there is no doubting the need for updating and revision, some conservationists have reservations about the punitive emphasis mooted by the implementers. In the UK, where there remains a lucrative egg-collecting and raptor black-market (mainly abroad), there are reasonable grounds for heavy fines. But £5000 for 'disturbance of Peregrines' or £2000 plus £4000 costs for grubbing up a hedge, albeit a medieval hedge: Is this really the direction in which we want conservation to go? In Ireland, where punishment for what are still widely perceived as rural pastimes or farming prerogatives smacks of 'gentry-law', punitive measures can be counter-productive.

It seems to me that conservation is really a 'hearts and minds' exercise. The optimism achieved along the Shannon was won by the slow process of building trust, resulting in cooperation from sectors with very different agendas. Certainly effective legislation is important but only in tandem with a continuing and widening educational programme that seeks to engender the reality that a thriving, diverse birdlife is a desirable aspiration for us all.

NOTES

The EC Directive on the Conservation of Natural Habitats and of Wild Flora and Fauna (usually referred to as the *Habitats Directive*) dates from 1992. It is part of the 'Natura 2000' programme. The core strategy is the establishment of a network of Special Areas of Conservation throughout Europe. This conservation designation recognises not only the importance of these areas for birds but also for biodiversity in general.

The *Berne Convention* which came into operation in 1982 obliges member states to protect their flora and fauna, particularly those that are endangered or vulnerable. It also obliges them to promote national policies for the conservation of wild flora and fauna and their habitats.

The *Bonn Convention* is concerned with migratory species. As regards our birds this Convention applies most particularly to our wintering waterfowl – ducks, geese and swans – some of which (the Greenland white-fronted Goose, for instance) are internationally rare.

International legislation which concerns our birdlife exists in the form of the *Ramsar Convention* which was signed in Iran in 1971. This legislation concerns the protection of internationally important wetlands, 21 of which are listed for the Republic. Its brief is not only the protection of the wetlands as entities but also their significance as habitats for birds. (For further information, see the *Red Data Book* 2: A. Whilde, 1993.)

72 *Ancient oaks, Glenstal*

73 *Fenland, Youghal*

RE-INTRODUCTION

The question of re-introducing birds that were once native here ought, one might think, to be greeted with at least passivity if not enthusiasm. Most naturalists are highly positive, seeing re-introduction as a redressing of an ecological deficit. It can be much more difficult to gain support from the general public. Since sectoral priorities are often both wide-ranging and specific, what might appear to be a 'good thing to do' is often opposed by a group with a conflicting agenda.

Some bird species are less contentious than others. It would be difficult to find trenchant opposition to the conservation efforts being carried out to save the Corncrake: this is partly as a consequence of its symbolic significance, particularly among country people. Bitterns and Cranes are gone so long (even the folklore of their former presence has all but vanished) that it is impossible to appeal to people on the same basis. The best we can reasonably aim for with these spectacular birds is to set aside sufficient habitat, wait optimistically for eventual natural recolonisation and hope that their visual appeal and benign behaviour will secure their permanent return.

Birds of prey are a different story. The fact that we have lost half of our native raptors means little to most people and might well elicit a 'good riddance' from some such as farmers, game managers and garden bird enthusiasts. Next to the Wolf, the eagle in its heyday must have been perceived by the livestock-tending community as the most undesirable predator in the countryside. Its extermination, while perhaps not brought about by farmers, was doubtless sanctioned by them. The Scottish experience has shown that this attitude is no longer justified. Golden Eagle predation on lambs lies well down the list – behind renegade dog, Fox and Raven – and, despite the hundreds of eagles breeding and hunting in the Highlands, the claims against eagles are nowadays few and far between. Long-term monitoring has shown that only the occasional, mainly sickly lamb is taken (RSPB, 1998) and the farming community is cognisant of this. The reputation gained by large raptors such as eagles in 18th- and early 19th-century Ireland and Scotland was for their brazen attacks on farmyard livestock – piglets, poultry etc. – which are no longer reared in open, unprotected farmyards. So, the circumstance of their revilement has to all intents and purposes evaporated. Even the managers of the grouse moors, who would have more grounds to be concerned, have learned to live with eagles. They do not cite eagle predation as a threat to the sustainability of stocks (O'Toole, 1999). It might be argued that the Golden Eagle case is not a fair one since, unlike Ireland, there has been unbroken occupancy of the Scottish Highlands so that relationships between bird and people have had time to evolve.

A better example would be the Red Kite, which, though it lingered a century longer in Scotland than in Ireland, also died out there. The RSPB reintroduction programme in two separate locations (in central and north Scotland) has produced some remarkable findings. Not only have the Kites been found to shun healthy game and livestock but they have been shown to be most effective carrion disposers, particularly of dead rabbits. In acting thus as effective eliminators of disease and pollution they have been welcomed by most of the farming community. They have also begun to nest in farmland, occupying moderately-sized,

74 *Blasket Isles – traditional White-tailed Eagle habitat*

mature stands of trees, demonstrating the developing symbiosis. The perception of the Red Kite as the shy denizen of the forest, which may have been the case in rural Ireland in medieval times, is not reflected in the manner of repopulation in Scotland, at any rate. They do continue to inhabit dense woodland in their native stronghold in Wales. As with the eagles, the cause of their odium – their propensity to swoop on young farmyard poultry – no longer exists because of changes in farming practice. On paper, therefore, the Red Kite would be a good candidate for a reintroduction programme, though, given the success in Scotland (and also in England where there have also been recent reintroductions) repopulation by natural dispersal may soon occur here. The question of perception remains: the behaviour of a large bird of prey hovering above sheep pasture can be easily misconstrued.

In the event of an active re-introduction plan being sanctioned, the most prohibitive factor initially might not be perception; it could well be finance. The reintroduction of the White-tailed Eagle on Rhum (or Rum) was enormously expensive. Over eighty Eagles were brought in from Norway to establish a viable breeding colony. The expense in first capturing the birds and then flying them in chartered planes and releasing them under controlled

supervision could only have been achieved by high levels of cooperation and an open account running to hundreds of thousands of pounds. The well-intentioned *ad hoc* efforts which have been tried to date in Ireland have almost no chance of success. As John Love has pointed out in his book *The return of the Sea Eagle* (1983) about the Rhum programme, a number of factors had to come together to promise success. Support was forthcoming from the most unlikely sources – the RAF for instance, who flew in young eagles from Norway, and the Eagle Star Insurance Company, who generously provided sponsorship. Wide-ranging cooperation of this kind will be the key to a similar programme here, assuming it gets the red light. There are those who feel that in time repopulation will occur naturally – from post-breeding dispersal of Scottish birds – and that 'nature' should be allowed to take its course. It would be nice to think that Ireland would provide a ready sanctuary for the expansion of the Scottish re-introductions. There is no guarantee, though, that the Scottish experiment will be an on-going success (though it looks very good at present) nor can we assume that sufficient numbers of eagles will drift to suitable habitat here to continue the success story. Again, one of the great imponderables is public response. How would the hundreds of people whose living depends on the harvest of the sea view the arrival of a large, fish-eating raptor with an appetite to rival that of the most voracious Cormorant? Probably sceptically, at least initially.

One point worth considering. The conditions which, even a century and a half ago supported resident Sea Eagles – unpolluted lakes and coasts, fish fecundity several times that of today – no longer exist. If a viable population of Sea Eagles is to survive, it might need a degree of 'sustenance intervention'. There are sufficient seabirds on the cliff and stack sites to act as an alternative to their primary food. (Fulmars, which were not on the cliffs before eagles became extinct, are often taken nowadays: they may well prove to be an important alternative food source for re-introduced eagles of both species.) But eventually, as is the case in Norway, Sea Eagles would gravitate to the fishing ports and new fish-farms for scavenging – an activity to which they are naturally predisposed. This is where human tolerance would be most tested. A compensation mechanism might be necessary initially, but Scottish experience has again shown that people earning their living along the coast have tended to welcome rather than scorn the Eagles. If sheep-farmers can be induced to abandon the use of poison and perhaps be presented with a mechanism for compensation, then there are reasonable prospects for a working relationship from that sector.

A cooperational/compensatory programme would be an even more important prerequisite to a re-introduction programme for the Golden Eagle, since it is more of a hunter than the Sea Eagle. It had been thought that, given the high density of breeding Golden Eagles in Scotland (422 pairs, RSPB, 1995) there was a good chance that a pair would soon attempt to nest here. Over the past twenty or so years, however, there has been no real increase in recorded sightings, and the chart of post-breeding dispersal (Watson, 1997) shows that young Golden Eagles (birds that might establish new territories) disperse *least* to the south-west – towards Ireland. The one-off breeding in Antrim forty years ago (which did not result in repopulation) came apparently from the Kintyre region, but since its afforestation in recent decades Golden Eagles have more or less vacated this peninsula. The evidence therefore points to the need for positive intervention to facilitate repopulation in Ireland. Even with the present 'wait and see' policy an advance programme among hill farmers and landowners in the remote areas where reintroduction might be considered or where eagles

might turn up anyway, would certainly be appropriate at this stage. This could take the form of a familiarisation programme carried out through a farm advisory body by an eagle expert. It would be a tragedy if even one Golden Eagle was lost through ignorance for the sake of inclusiveness.

(A raptor group has been established whose aim is to re-introduce Golden Eagles to Ireland. The plan is a co-operative venture between BirdWatch Ireland, the Irish Raptor Study Group and *Dúchas*, the National Parks and Wildlife Service. The programme recommends the release of 60–80 Scottish Golden Eagle chicks into a location in Ulster, over a five-year period. Costs are of the order of £35,000 per annum, not a prohibitive sum for so daring a project. Already a considerable level of momentum has been generated both here and in Scotland, and an outline report has been drawn up by Lorcan O'Toole and [the late] Jim Haine.) While understandably, many would favour 'natural' over man-induced repopulation of Golden Eagles, there is a much better chance of success using the 'Rhum method'. A century has passed without natural repopulation (excepting the Fair Head breeding). Perhaps the time is right to redress at least this imbalance through positive intervention.

What a spectacle it would be to see these great birds once again soaring over mountains in Ireland as they did up until a century ago. The next generation who would benefit from this spectacle would be more concerned about conserving the eagles than about how they returned.

Monitoring has shown that a substantial percentage of Scottish Ospreys disperse to and through Ireland. Single birds are regularly seen on suitable breeding habitat in spring, and in recent years pairs have been seen. One or two pairs have even spent the summer in suitable breeding habitat both in the west and in the north of Ireland. The likelihood of breeding here is strong, and positive intervention in the form of 'Rhum-style' introduction is surely unnecessary. The Irish Raptor Group, acutely aware of imminent repopulation, have been erecting simple 'Osprey platforms' in likely breeding sites. These makeshift supports – a few stout branches secured in a rough square near the top of a tall tree – are

75 *Golden Eagle eyrie site, Donegal*

76 *Osprey Island, Killarney lakes*

77 *Osprey platform*

not nest substitutes but rather 'inducement scaffolding' for would-be nesters. They have been shown to work effectively in Scotland and will surely also work here, in time. Already on several lake sites, in several Irish counties, Osprey platforms await the return of the tonsured fishermen.

The failure of the reintroduction of the Capercaillie in Ireland in the 19th century was seen by Deane (1979) as evidence that the Capercaillie had never been indigenous in Ireland. (In fairness he knew nothing at the time of writing about the confirmatory archaeo-zoological evidence which has since come to light.) The Irish programme was obviously as a result of the success which, since 1837, had been achieved in Scotland. The 'three brace' that were sent to Lord Bantry in Glengariff, from Sweden, in about 1842 survived for less than a year (*per* Thompson). According to reports they succumbed, one by one, to an unidentified ailment. A similar fate befell Capercaillie hatchlings at Lord Orkney's estate in Buckinghamshire at about the same time. Colonel Cooper tried again at Markree Castle estate in Co. Sligo (1876–7) – again unsuccessfully. There may have been other attempts at reintroduction besides.

It would appear that the Irish attempts were jinxed as much by the half-heartedness of the efforts as by the unpredicted afflictions which killed off the stock. However, there is no doubt that the successful reintroduction into Perthshire was effected due to the survival of Scots pine forest into which the Swedish birds readily adapted.

Deane made the point that with the availability of conifer plantation nowadays successful introduction may yet occur in Ireland and there seems to be an inevitability about this in view of the fall off in the numbers of our other game birds.

If Capercaillies are to be brought back – for hunting or simply to re-enrich our countryside – I would like to see it done in the context of habitat reconstruction. In other words it would be better, in my view, to reconstruct a Capercaillie-friendly area, with Scots Pine heath, rather than simply turning the birds loose into Sitka Spruce plantation where they might or might not survive. The project would involve reinstating a substantial Scots Pine upland region, say in Connemara or adjacent to the Burren in Co. Clare where pine cover is known to have existed into historical times. It could be laid out as an extension to one or other of the existing national parks. A semi-natural conifer forest of this kind would have conservation value in its own right and would act as ready repository for other conifer-friendly fauna including birds such as Capercaillies, Goshawks, woodpeckers, Crossbills etc. – whether directly introduced or not.

The issue of habitat is fundamental to all introductions. While we are focusing here on birds, we are not viewing them in isolation: we are also concerned with context. Detached from a viable, self-sustaining environment the concept of reintroduction is merely hypothetical.

With each passing year more and more hardwoods are being planted. Soon Ireland will be growing sufficient oak, ash and beech for domestic requirements. Coillte and other forestry companies are playing a significant part. Other organisations like the Tree Council of Ireland and Crann are at the forefront in encouraging urban forestry, the planting of green space in cities, north and south, for amenity and wildlife purposes. All this augurs well for the future of some of our lost birds of prey, assuming their eventual return. We are, however, only beginning to redress an environmental dearth which reached an all-time low in the 1950 (before the spread of commercial forestry) when Ireland's woodland cover, at about 4%, was the lowest in Europe. This contrasts markedly with the estimated 12.5% cover towards the end of the Gaelic period, prior to the final Jacobean clearances.

Regardless of the requirements of our large raptors, I believe that there is a case for woodland reconstruction *per se*. One or more extensive hardwood forests could be reinstated where records tell us they once were, not for any particular reason (avian or otherwise) but as an environment enrichment project in its own right. We know that oak woods, tens of thousands of acres in extent, remained in Ireland up until four hundred years ago. Two well documented examples were the Glen of Aherlow in Munster and Glenconkeyne in Ulster. But there were others (McCracken, 1971). It is not difficult to locate such places either from old documents or through placenames. The project would require a high degree of planning, organisation and cooperation. However, once under way the potential, in terms of timber supply, recreation and amenity, would be obvious. Given the present difficulties in farming, an enterprise of this kind, with high traditional values, could well elicit considerable support from that quarter. It would have the important effect of allowing continuity in farming activity that might otherwise be forced to disintegrate. If the pilot scheme was of a sufficient scale (say, 1000 hectares minimum), the wildlife potential would be enormous too. Not only would a woodland of this size prove irresistible to most large raptors but it would offer release opportunities for a selection of extinct mammals besides. In overall scientific terms its potential as a national woodland gene-pool could not be overstated.

With wetlands, circumstances are somewhat different but again the question of size is highly pertinent. In Ireland we still have many small and marginal fens but only a handful are greater than 100 ha. – Pollardstown in Co. Kildare, Ballyvergan near Youghal, Akeragh in Kerry, and a few others. Birds such as Marsh Harrier and Bittern are mainly found in extensive wetlands where territorial and hunting requirements exist. Perhaps it would be possible – using the same arguments and methods as above – to buy up and develop a few extensive fens and reed swamps like those which existed a century or so ago. Large tracts of bogland, albeit cut-away bogland, are still found in the Irish midlands.

Bord na Móna have, of late, recognised the amenity and wildlife potential of these areas. A pilot scheme at Boora bog in Co. Offaly has converted several hundred acres of 'wasteland' into a considerable tourist attraction. A section at Tirraun has returned to fen and lake and already over 150 bird species have been identified from the cleverly designed walkways and hides. A similar plan has been outlined for Littleton in Tipperary. Over a ten-year period it is planned to return a network of midland cut-aways to lake and fen systems. The process involves utilising calcareous glacio-fluvial deposits from beneath the bog to ensure that the water and hence the ecology of the new wetlands is base-rich and productive. The net result will be a mosaic of more than a thousand hectares of new wetland surrounded by worked out and abandoned bogland. Ultimately this will become a Mecca for waterfowl – particularly in winter. It may well also fulfil the habitat requirements for the return during the nesting season of some of our spectacular but still absent wetland birds. Already the Marsh Harrier has been seen gliding over the emergent vegetation at Tirraun. Who is to say that a few decades from now the midlands bogs won't once again ring to the bell-notes of Cranes as they did in the Middle Ages.

To sum up, the provision of habitat is likely to attract at least a few of our lost birds of *their own volition*. Those that come to mind are the birds of prey – the Red Kite, the Buzzard, the Goshawk and the Marsh Harrier – all of which are doing well in Britain. The Bittern and the Crane are less likely returnees – at least in the short term. It is difficult to see how sedentary species such as the Capercaillie and the Great-spotted Woodpecker could repopulate without positive intervention, even with the reinstatement of suitable woodland. I see nothing wrong with such a measure and would advocate it in the event of the hardwood forest project which I have proposed above becoming a reality. It is difficult to know what to do for the Corncrake and the other vanishing game birds, over and above what is being done. Perhaps time will prove us all wrong and they will stage their own recovery, demonstrating yet again the intrinsic resilience of nature. But that does not seem likely at the present time. We will unfortunately always have the example of the extinct Great Auk to warn us that, no matter how successful we are in our efforts to create a richer environment, we cannot recreate the past.

CONCLUSION

This conclusion has been written more than once. Initially, twenty years ago, it was a trenchant attack on the 'perpetrators' of wetland drainage, the forest exploiters, the instigators of agricultural intensification … those sectoral interests which I deemed to have contributed to the disappearance of the lost birds. There is still a bit of that – it is difficult, knowing the history, not to be resentful about what might have been – but, mercifully, my attitude has mellowed with the maturing of the book. The process of maturation has resulted in condensation. A substantial volume of background material, formerly included, has been jettisoned in the revision.

My emotional response was not devoid of a *rationale*. I had just emerged from Trinity, a Master's in Environmental Science under my belt and indignation, born of the subject of my thesis (turlough drainage in Galway), in my gut. My study had revealed serious environmental destruction resulting in unquantifiable loss to wildlife and questionable benefits to a small number of farmers. But these were uncomplicated times. The Common Agricultural Policy was the unchallenged code, and the idea of the countryside being anything other than the testing ground for agriculture was strictly for the birds. We have moved on since then, and the increasing uptake in REPS shows that attitudes towards the sustainability of our countryside resource has dramatically advanced. The Environmental Impact Statement has become mandatory and the Arterial Drainage programme has ground almost to a halt.

But why all this reasonableness? The lost birds remain lost, and the exploitation and short-sightedness of our predecessors is still to blame, isn't it? If we look dispassionately at the facts, there is certainly good historical reason to implicate certain sectoral interests. Our eagles do appear to have been persecuted by uncompromising sheep-farmers and gamekeepers and finished off by trophy collectors. The Goshawk and the Capercaillie were surely victims of the final destruction of their forest habitat for the sake of the timber industry. The Great Auk was hunted to oblivion by seafarers for food. However, if we look more closely at the historical facts we learn also that the Great Auk's demise was hastened by an earthquake, which wiped out a relict colony in Iceland; nor do we now know how significant a factor earthquakes were overall, in its extinction.

A critical look at the narrative of disappearance of other lost birds points to the influence of contributory factors. The simplistic explanation is, in most of the case studies, blatantly unsatisfactory. The Corncrake story illustrates this clearly. The popular perception is that modern agricultural practice has been responsible for the disappearance of this bird. Most naturalists agree that its decline has been *hastened* thus, but others question whether or not it was the instigative factor. The decline was noted in parts of Europe decades before mechanised mowing was commonplace, before agricultural intensification was widespread, before crop-sprays and bulldozers began to impact.

And large tracts of Corncrake-friendly habitat remain in Ireland – in the Burren (which has been 'Corncrakeless' for perhaps thirty years), in island meadowland, in coastal machair etc. So how culpable is modern agriculture? Stephen Mills in his excellent *Nature in its place* (1987) points to the decline in the diversity of invertebrate food, a consequence of the use of pesticides and other chemicals, as a prime factor in the Corncrake's decline. Later, however, he surmises that an unidentified 'biological syndrome' might be involved. Ireland's geographical position – on the western fringe of Europe has surely had a bearing – as it has in the other lost or declining species, but some would argue that this is of no more than secondary significance. Ireland's peripherality certainly makes it more difficult for lost birds to return given that their centres of population are for the most part on continental Europe.

The Corncrake has, for millennia, been subject to increasing pressures on its extraordinary annual migration to south and south-east Africa. While research has failed to reveal factors which might account for the decline in the African winter quarters, can we be equally relaxed about the question of migration? Could the inexorable southward march of the Sahara, for instance, present an increasing obstacle to the leg-dangling rail? What about the thousands that are killed annually for food in the Mediterranean countries which have traditionally hunted European migrants? And what about the less quantifiable factors which tend to be sidelined for that very reason? Are we to ignore the possible effects of climatic change. Global warming could hardly have caused a shift in the Corncrake's population base, could it?

And perhaps more speculatively, is there an evolutionary case to be answered as regards the longevity of this archaic bird? Could it have passed its 'ascendancy phase', and like the California Condor be sliding towards a natural 'bowing out' – like the plethora of other birds which have come and gone in the 150 or so million years since birds began. We tend to think that, all other factors being equal (food, habitat, predator/prey balance etc.), a bird or other organism will thrive, indefinitely. We readily compartmentalise, for analysis' sake, and come up with models (some of which have helped with our understanding) based on our current perceptions. But how well equipped are we, in fact, to consider the vastness of the evolutionary timeframe?

So, what is this text really about? Is it simply a dispassionate reflection on a dozen or so birds no longer found here? Is it a vehicle for ornithological nostalgia, or sectoral blame? Yes, nostalgia and culpability have figured, on a personal basis, but

there is also a measure of genuine regret that the lost birds are no longer around for all of us to enjoy.

However, the book is really about a greater regret: it is about lost plants – vanished ruderals, Corncockle and Darnel – which, until a generation or two ago were as commonplace as the Corncrake in the country grainfields. It is about our lost animals – Wild boar, Wild Cat and Wolf – unaccommodated inhabitants of the wildwood. It is about the small, the inconspicuous and insignificant flora and fauna which have departed the scene before being recognised and celebrated by modern nature-lovers.

In the broad context this book is about lost forests, marshes, boglands – pristine habitats swallowed up or converted to something else in the name of development. It is, in short, about regret for the loss of natural wealth.

At a time when we, the Irish, are enjoying wonderful economic benefits we are undoubtedly selling ourselves short elsewhere. We are, I believe, sacrificing profound values (like environmental appreciation) for a more immediate, shallower set. There is much evidence for this, but perhaps the decline of media interest in celebratory, as distinct from contentious and controversial environmental themes, reflects this trend.

While I was at Trinity, a visiting lecturer steeped in the logic of the cost/benefit system (more in vogue then than now) tried to sell the class the idea that a Greenland White-fronted Goose had a monetary value based on conservation costs, gun licences and other assessable considerations. The national 'flock', he estimated, was worth some hundreds of thousands of pounds. I was having none of this: something inside me baulked at the idea of reducing a wonderful wild bird to an item on a price list. I was glad to see that most of the class took a similar view. How could the total value of a living thing be quantified in such crass terms? This piece of madness brought home to me, more forcefully than anything I have come upon since, the urgent need to explore the re-evaluation of the priceless.

In losing our birds we are losing more than simply birds: we are losing cultural symbols. This loss is also the loss of meaning – in science, in literature, artefact and music, in natural texture and colour. It is a diminishment of that which we hold dear. Nor has the arrival of the new replaced the lost. If comparisons are acceptable, recent 'blow-ins' such as Collared Doves and egrets, exotic and all as they are, are no substitute for species such as eagles, whether in ecological terms or for their cultural value.

While it is not so noticeable in our birds, some of our wildlife blow-ins have proved to be far from benign. As elsewhere, we have experienced the disruption of immigrants which often threaten the stability of native or endemic species. Island histories throughout the world are replete with examples of the devastation wrought by virulent introduced predators. In Ireland the effects are difficult to isolate due to the long history of human migration to, from and through the island. The fact that a significant fraction of our present wildlife has come to us by the direct or indirect intervention by man does not mean that we should admit every organism which appears on our shores. We are in the throes of learning to cope with the latest batch – the American Mink, the Zebra Mussel and the New Zealand Flatworm.

The question as to *why* we should favour the native over the introduced, in philosophical terms, can be difficult to articulate. Some would argue for 'the more the merrier' or 'the more exotic and colourful the better'. I would disagree, despite finding the issue difficult to reconcile rationally. I am drawn towards the notion of 'quota' – the suite of organisms which, by virtue of geology, geographic location, climate, etc. has, over the immensity of time, established itself here. This imbues our native wildlife with a sense of *specialness* or *preciousness* which I find easy to live with. The argument would be neater if we could leave man out of the equation. The quota has become established, however, despite and because of man's influence. However more convenient it might be, were it otherwise, we have also to live with this reality.

As far as I am concerned, the continuation of a free-for-all on the one hand, or a managed strategy approach on the other, offer equally unsatisfactory options for the future. The first liberates us from any responsibility for our natural circumstances; the second shapes us into anxious custodians of our natural heritage. It is difficult to imagine a more onerous choice. Which way do we go? Whatever we decide we can no longer plead ignorance.

Ultimately it will surely be a question of balance. In pre-industrial times (before 1600, say) nature and humanity combined, decided the balance. Nowadays it is the pulsing and consensus of human affairs – increasingly sanitised and detached from nature – which decides.

The birds of the case studies which are the substance of this book were once native 'everyday' birds. We can no longer look upon them, however, as our predecessors once did: a chevron of Cranes trumpeting over a raised bog; a cock Capercaillie strutting its stuff from the depth of a pine forest; an Osprey struggling to take off from a western lake with a salmon in its talons; a Corncrake calling monotonously from the meadow beyond the haggard. There is a temptation to apply our modern value system to the past, but it must be avoided. The environmental criteria of the Middle Ages, for instance, had little in common with those of today: birds were food, commodities or vermin. Nevertheless, we come across snippets which suggest surprising affection for birds (in the Early Irish poetry) and even a sense of protectiveness, albeit for the sake of future availability. Nor should we lose sight of the fact that man's earliest expressions of wonderment took the form of drawings of wildlife on cave walls. Perhaps we have undervalued the cultural sense of our predecessors. Perhaps nature appreciation has evolved less than we think.

I no longer see myself primarily as an eco-warrior. Others will take on that role and focus public opinion on environmental imbalances that should not, and need not be. I prefer an undramatic, yet optimistic look forward, to a gradual winning-over of public opinion. To an Ireland once again replete with natural diversity. The return of nine out of the eleven lost birds listed is quite possible, in my view. Hopefully this book will highlight that and help us distinguish between the probable, the improbable and the gone, never to return. I would like to think that a project of this kind would have positive cultural and social ramifications, no matter how small.

Some of the most enduring words about lost birds have come from North America. *The last of the Curlews* by Fred Bodsworth (1954) sadly but delightfully tracks the migration of the last pair of

Eskimo curlews from their Canadian breeding grounds to Patagonia and back. The book has enjoyed enormous popular appeal, has been continuously in print and has recently been made into an animated film. American people regret the passing of what once was one of the most abundant of American birds. The historical inserts in Bodsworth's book indicate that this bird was shot, almost into oblivion, by settlers who valued it simply as food.

Another American bird which went the same way was even more remarkable. The Passenger Pigeon may have been the commonest bird in the world in the 1700s. It migrated in countless millions through north-east America, more or less in a single flock that took days to pass, devouring enormous quantities of seed and grain *en passant*. It was thus simple to kill, and a century of concentrated hunting brought about its total extermination. This is what Alo Leopold had to say about the Passenger Pigeon:

> We are told by economic moralists that to mourn a pigeon is mere nostalgia; that if the pigeoners had not done away with him, the farmers would ultimately have been obliged, in self-defence, to do so.

This is one of those peculiar truths that is valid, but not for the reasons alleged.

The pigeon was a biological storm. He was the lightning that played between two opposing poles of intolerable intensity: the fat of the land and the oxygen of the air. Yearly the feathered tempest roared up, down and across the continent, sucking up the laden fruits of the forest and the prairie, burning them in a travelling blast of life. Like any other chain reaction, the pigeon could withstand no diminution of his own furious intensity. When the pigeoners subtracted from his numbers, and the pioneers chopped gaps in the continuity of his fuel, his flame guttered out with hardly a splutter or even a wisp of smoke.

Today the oaks still flaunt their burden at the sky, but the feathered lightning is no more. Worm and weevil must now perform slowly and silently the biological task that once drew thunder from the firmament.'

[from *A Sand County Almanac*, 1949]

SPECIFIC NAMES OF BIRDS IN TEXT

Red-throated Diver   *Gavia stellata*
Black-necked Grebe   *Podiceps nigricollis*
Fulmar   *Fulmarus glacialis*
Cormorant   *Phalacrocorax carbo*
White Pelican   *Pelecanus onocrotalus*
Bittern   *Botaurus stellaris*
Little Egret   *Egretta garzetta*
Grey Heron   *Ardea cinerea*
White Stork   *Ciconia ciconia*
Mute Swan   *Cygnus olor*
Bewick's Swan   *Cygnus columbianus*
Whooper Swan   *Cygnus cygnus*
Bean Goose   *Anser fabalis*
White-fronted Goose (Greenland)   *Anser albifrons flavirostris*
Greylag Goose   *Anser anser*
Canada Goose   *Branta canadensis*
Teal   *Anas crecca*
Mallard   *Anas platyrhynchos*
Labrador Duck   *Camptorhynchus labradorius*
Goosander   *Mergus merganser*
Ruddy duck   *Oxyura jamaicensis*
California Condor   *Gymnogyps californianus*
Black Kite   *Milvus migrans*
Red Kite   *Milvus milvus*
White-tailed Eagle   *Haliaetus albicilla*
Marsh Harrier   *Circus aeruginosus*
Hen Harrier   *Circus cyaneus*
Goshawk   *Accipiter gentilis*
Sparrowhawk   *Accipiter nisus*

Buzzard   *Buteo buteo*
Golden Eagle   *Aquila chrysaetos*
Osprey   *Pandion haliaetus*
Kestrel   *Falco tinnunculus*
Merlin   *Falco columbarius*
Hobby   *Falco subbuteo*
Peregrine   *Falco peregrinus*
Red Grouse   *Lagopus lagopus*
Ptarmigan   *Lagopus mutus*
Black Grouse   *Tetrao tetrix*
Capercaillie   *Tetrao urogallus*
Grey Partridge   *Perdix perdix*
Red-legged Partridge   *Alectoris rufa*
Quail   *Coturnix coturnix*
Andalusian Hemipode   *Turnix sylvatica*
Water Rail   *Rallus aquaticus*
Spotted Crake   *Porzana porzana*
Corncrake   *Crex crex*
Moorhen   *Gallinula chloropus*
Takahe   *Notornis mantelli*
Coot   *Fulica atra*
Crane   *Grus grus*
Sandhill Crane   *Grus canadensis*
Great Bustard   *Otis tarda*
Stone Curlew   *Burhinus oedicnemus*
Golden Plover   *Pluvialus apricaria*
Dotterel   *Charadrius morinellus*
Lapwing   *Vanellus vanellus*
Dunlin   *Calidris alpina*

Snipe   *Gallinago gallinago*
Woodcock   *Scolopax rusticola*
Black-tailed Godwit   *Limosa limosa*
Eskimo Curlew   *Numenius borealis*
Red-necked Phalarope   *Phalaropus lobatus*
Roseate Tern   *Sterna dougalli*
Little Tern   *Sterna albifrons*
Guillemot   *Uria aalge*
Razorbill   *Alca torda*
Great Auk   *Pinguinus impennis*
Puffin   *Fratercula arctica*
Stock Dove   *Columba oenas*
Woodpigeon   *Columba palumbus*
Passenger Pigeon   *Ectopistes migratorius*
Collared Dove   *Streptopelia decaocto*
Carolina Parakeet   *Conuropis carolinensis*
Barn Owl   *Tyto alba*
Eagle Owl   *Bubo bubo*
Nightjar   *Caprimulgus europaeus*
Wryneck   *Jynx torquilla*
Green Woodpecker   *Picus viridis*
Great-spotted Woodpecker   *Dendrocopus major*
Lesser-spotted Woodpecker   *Dendrocopus minor*
Woodlark   *Lullula arborea*
Skylark   *Alauda arvensis*
Sand Martin   *Riparia paludicola*
Swallow   *Hirundo rustica*
Tree Pipit   *Anthus trivialis*
Yellow Wagtail   *Motacilla flava*
Grey Wagtail   *Motacilla cinerea*
Dipper   *Cinclus cinclus*
Wren   *Troglodytes troglodytes*
Nightingale   *Luscinia megarhynchos*
Redstart   *Phoenicurus phoenicurus*
Ring Ouzel   *Turdus torquatus*
Blackbird   *Turdus merula*

Song Thrush   *Turdus philomelos*
Mistle Thrush   *Turdus viscivorus*
Cetti's Warbler   *Cettia cetti*
Sedge Warbler   *Acrocephalus schoenobaenus*
Reed Warbler   *Acrocephalus scirpaceus*
Whitethroat   *Sylvia communis*
Pied Flycatcher   *Ficedula hypoleuca*
Marsh Tit   *Parus palustris*
Willow Tit   *Parus montanus*
Crested Tit   *Parus cristatus*
Coal Tit   *Parus ater*
Bearded Tit   *Panurus biarmicus*
Dartford Warbler   *Sylvia undata*
Whitethroat   *Sylvia communis*
Goldcrest   *Regulus regulus*
Nuthatch   *Sitta europaea*
Treecreeper   *Certhia familiaris*
Red-backed Shrike   *Lanius collurio*
Jay   *Garrulus glandarius*
Magpie   *Pica pica*
Chough   *Pyrrhocorax pyrrhocorax*
Hooded Crow   *Corvus corone*
Raven   *Corvus corax*
Starling   *Sturnus vulgaris*
Tree Sparrow   *Passer montanus*
Goldfinch   *Carduelis carduelis*
Siskin   *Carduelis spinus*
Twite   *Carduelis flavirostris*
Linnet   *Carduelis cannabina*
Crossbills   *Loxia leucoptera*
Bullfinch   *Pyrrhula pyrrhula*
Hawfinch   *Coccothraustes coccothraustes*
Snow Bunting   *Plectrophenax nivalis*
Yellowhammer   *Emberiza citrinella*
Cirl Bunting   *Emberiza cirlus*
Corn Bunting   *Miliaria calandra*

BIBLIOGRAPHY

MANUSCRIPT SOURCES

Molyneux, W., Papers on the natural history of Ireland 1684 etc. Trinity College Library, Dublin

Molyneux, Gwithers et al. on the natural history of Ireland, 18th cent., 2 vols. (from communications to the Dublin Society): MS 883, National Library, Dublin.

The *Zoilomastix* of Philip O'Sullivan Beare, *c.*1625: MSS 2759–62, National Library, Dublin.

Plunkett, Richard: Irish/English dictionary, 1662, Marsh's Library, Dublin.

Henry, (Revd W.): Hints towards a natural and topographical history of the counties of Sligo, Donegal, Fermanagh and Lough Erne, 1739: MS 2533, National Archives, Dublin.

Pococke, Dr, Tour of Ireland, 1752: MS 887, Trinity College Library, Dublin.

A description of Ireland, 1598 (from a Ms made in 1790 for Revd J. Beatty): MS 24 H.15, Royal Irish Academy.

John Templeton, Diaries, 1806–1825 (microfiche), Ulster Museum, Belfast.

Ordnance Survey Memoirs 1830–32 (various authors), Royal Irish Academy, Dublin.

Donatus, Sister, 'Beasts and birds in the Lives of the early Irish saints' (M.A. thesis, 1934), Trinity College, Dublin.

Parker-Hutchinson: Notes on natural history, 1870–73 and Notes (and catalogue) of the birds of Lough Derg and the riding of North Tipperary, 1872: MS 1205, National Library, Dublin.

Knox, H.B., Notes on the natural history, 1861–66: MS 4461, National Library, Dublin.

Anon., Early illustration of Great Auk, 18th cent.?: British Library, MS 27362 (Add.) Folio 42, O.N. 8552 612.

Galway Corporation Book A, (1638): f. 170, Illustration of game bird.

Book of Kells (9th cent.), four symbols page: f. 290v, Eagle of St John illustration. Trinity College, Dublin.

Ó hAonghusa, Amhlaoibh, 'Irish names of birds' (unpublished collection) (*c.*1940).

BOOKS

Allingham, H. (1879) *Ballyshannon. Its history and antiquities.* Londonderry.

An Roinn Oideachais (1978) *Ainmneacha plándai agus ainmhithe.* Dublin.

Armstrong, E.A. (1946) *Birds of the grey wind.* London.

—— (1958) *The folklore of birds.* London.

—— (1973) *St Francis: Nature mystic.* London.

Badlock, D. (1984) *Wetland drainage in Europe.* Nottingham.

Bannerman, D. A. (1956) *The birds of the British Isles*, vol. 12 (containing 'A short comparative list of Celtic bird names' by W. Nicolaisen), Edinburgh.

Bardon, J. (1992) *A history of Ulster.* Belfast.

Barrington, R.M. (1900) *Migrations of birds as observed at Irish lighthouses and lightships.* Dublin & London.

Batten, L.A., Bibby, C.J., Clement, P., Elliot, G.D., Porter, R.F., (eds.) (1990). *Red Data Birds in Britain.* London.

Baxter, E.V. & Rintoul, L.J. (1953) *The birds of Scotland.* Edinburgh.

Bedel, W. (1685–90) *Translations of biblical texts.* London.

Benson, C.W. (1901) *Our Irish songbirds.* Dublin.

Berkenhout, J. (1789) *Synopsis of the natural history of Great Britain and Ireland.* London.

Bijleveld, M. (1974) *Birds of prey in Europe.* London.

Bodsworth, F. (1954) *The last of the Curlews.* London.

Boethius, H. (1526) *History of Scotia.*

Botius, G. (1652) *Boate's natural history of Ireland.* London.

Brewer, J.S. & Bullen, W. (1867) *Calendar of Carew manuscripts.* London.

Brown, L. (1976) *British birds of prey.* London.

Bruton, R. & Convery, F.J. (1982) *Land drainage policy in Ireland.* Dublin.

B.T.O. & I.W.C. (1976) *The atlas of breeding birds in Britain and Ireland.* Tring, U.K.

Buffon, Comte de (1793) *The natural history of birds,* 9 vols. London.

Burton, J. F. (1995) *Birds and climate change.* London.

Cabot, D. (1999) *Ireland* (New Naturalist Series). London.

Clark, W. (1971) *Rathlin – disputed island.* Waterford.

Colles, R. (1919) *The history of Ulster,* vol. 2. London.

Cohen, A. & Serjeantson, D. (1996, rev.) *Bird bones from archaeological sites.* London.

Council of Europe (1984) *Birds in need of special protection in Europe.* Strasbourg.

Coward, T.A. (1920) *Birds of the British Isles.* London.

Davis, S.J.M. (1987) *The archaeology of animals.* London.

Deane, C.D. (1954) *The handbook of the birds of Northern Ireland.* Belfast.

—— (1983) *The Ulster Countryside.* Belfast.

Derricke, J. (1581) *The image of Ireland,* ed. J. Small. (1883)

Dinneen, P.S. (1927) *Foclóir Gaedilge agus Bearla. An Irish/English dictionary.* Dublin.

Dixon, C. (1893) *Game birds of the British Islands.* London.

Dunaire Finn (12th cent.) *The book of the lays of Fionn* (transl. E. MacNeill, 1908). London.

Dubourdieu, J. (1802) *Statistical survey of the county of Down.* Dublin.

—— (1812) *Statistical survey of the county of Antrim.* Dublin.

Eager, A.R. (1980) *A guide to Irish bibliographical material.* London.

Edlin, H.L. (1956) *Trees, woods and man.* London.

Fairley, J.S. (1972) *Irish wild mammals. A guide to the literature.* Galway.

—— (1975) *An Irish beast book.* Belfast.

Falkiner, C.L. (1904) *Illustrations of Irish history & topography* (mainly 17th cent.). London.

Faragher D.C. (1979) *English-Manx dictionary.* Douglas.

Feehan, J. (1983) *Laois, an environmental history.* Portlaoise.

Fisher, J., Simon, N., Vincent, J., (1969) *The Red Data Book (Wildlife in danger).* London.

Forbes, A.R. (1905) *Gaelic names of beasts, birds etc.* Edinburgh.

Foster, J.W. (1997) *Nature in Ireland.* Dublin.

Fox, A.W. (1924) *Haunts of the eagle.* London.

Fox, J. (1988) *Irish birds.* London.

Fuller, R.J. (1982) *Bird habitats in Britain.* London.

Gardiner, L. (1923) *Rare, vanishing & lost British birds* (compiled from the notes of W.H. Hudson) London & Toronto.

Garrad, L.S. (1978) *Man and the environment in the Isle of Man.*

Goldsmith, O. (1769) *History of animated Nature.* London.

—— (1770) *The deserted village.* Dublin.

Gooders, J. (1983) *Birds that came back.* London.

Gosling, P. (1993) *From Dun Delca to Dundalk.* Dundalk.

Greenoak, F. (1979) *All the birds of the air.* London.

Grieve, S. (1885) *The Great Auk.* London.

Gurney, J.H. & Witherby, H.F. (1921) *Early annals of ornithology.* London.

Hall, R. (1984) *The Viking dig.* London.

Hagemeijer, J.M. & Blair, M.J. (1997) *The EBCC atlas of European breeding birds.* London.

Harris, W. (1774) *The antient and present state of the county of Down.* Dublin.

Harting, J.E. (1906) *Recreations of a naturalist.* London.

Harvey, W.H., Humphreys, J.D. & Power, T. (1845) *Contributions towards a fauna and flora of the county of Cork* (Cuverian Society). Cork.

Hawksworth, D.L. (ed.) (1974) *The changing flora and fauna of Britain.* London.

Heathcote, N. (1900) *St Kilda.* London.

Hoare, R.S. (1859) *A tour of Ireland* (1807). London.

Holinshed, R. (1586) *Holinshed's Irish Chronicle* (includes Stanihurst's 'Description of Ireland') London.

Hudson, W.H. (1895) *British birds.*

—— (1913) *Adventures among birds.* London.

Hutchinson C. (1979) *Ireland's wetlands and their birds* (IWC). Dublin.

Institution of Terrestrial Ecology (1982) *Predatory birds, pesticides and pollution* (National Environment Research Council).

Irish Times (1983–99) 'Where's That', 'Another Life' etc. Dublin.

Jackson, K. (1935) *Studies in Early Celtic nature poetry.* Cambridge.

Jardine, Sir W. (1838–42) *Birds of Great Britian and Ireland.* London

Joyce, P.W. (1901–2) *Irish names of places,* 2 vols.

—— (1906) *A smaller social history of ancient Ireland.* Dublin.

Kelly, F. (1997) *Early Irish farming* (Dublin Inst. of Advanced Studies). Dublin.

Kennedy, P.G., Ruttledge, R.F., & Scroope, C.F. (1954) *The birds of Ireland*. Edinburgh.

K'Eogh, Revd J. (1735) *Zoologica Medicalis Hiberniae*. Dublin.

Lack D. (1971) *Ecological isolation in birds*. London.

Leopold, A. (1949) *A Sand County almanac*. New York.

Lockwood, W.B. (1984) *The Oxford book of British bird names*. Oxford.

Love, J. (1984) *The return of the Sea Eagle*. Cambridge.

MacArthur, R.H. & Wilson, E.O. (1967) *The theory of island biogeography*. Princeton.

Mac Giollagunna, C.B. (early 18th cent.) *An Bonnan Buí* (transl. by T. MacDonagh). Dublin.

Mac Giollarnáth, S. (1940) *Feilire na Nean*. Dublin.

—— (1942) *Handbook of Irish Folklore*. Dublin.

McCarthy, C. (1900) *The game laws of Ireland*. Dublin.

McCracken, E. (1971) *Irish woods since Tudor times*. Newtownabbot.

McEvedy & Jones (1978) *Atlas of world population history*. London.

McKee, M. (1983) *James Sheals (naturalist & taxidermist)*. Belfast.

McLysaght, E. (1939) *Irish life in the seventeenth century*. Cork.

McSkimmin, S. (1811) *The history and antiquities of the county and town of Carrickfergus*. Belfast.

Marshall, J.D. (1836) *The natural history of the island of Rathlin*. Dublin.

Martin M. (1714; reprinted 1976) *Description of the Western Isles of Scotland*. Edinburgh.

Mason W.S. (1814–19) *A statistical account or parochial survey of Ireland*, 3 vols. Dublin.

Maxwell, N.H. (1838) *Wild sports of the West*. London.

Meyburg, B.U. & Chancellor, R.D. (eds.) (1994) *Raptor conservation today*.

Mhic Dhomhnaill, A. (1967) *Fealsunacht* (1842) ed. by C. Becket. Dublin.

Mills, S. (1987) *Nature in its place*. London.

Mitchell, F. (1986) *Shell Guide to reading the Irish landscape*. Dublin

Moffat, C.B. (1898) *The life and letters of Alexander Goodman More*. Dublin.

Moriarty, C. (1967) *A guide to Irish birds*. Cork.

Moryson, F. (1735) *A description of Ireland (c.1600)* Dublin.

Murton, R.K. (1971) *Man and birds*. London

National Association of Regional Game Councils (1986) *Irish shooting companion* ed. by J. Dunne and P. Lawlor). Galway.

Neeson, E. (1991) *A history of Irish forestry*. Dublin.

Nelson, E.C. & Walsh W.F. (1993) *Trees of Ireland*. Dublin.

O'Ceallaigh, T. (1925) *Saoghal Éanacha*. Dublin.

O'Donnell, T.J. (1960) *Selections from the Zoilomastix of Philip O'Sullivan Beare*. Dublin.

O'Donovan et al. (1835) Ordnance Survey letters. Dublin.

O'Flagherty, R. (1684) *A chorographical description of West or h–Iar Connaught*. ed. by J. Hardiman for the Irish Archaeological Society, 1846. Dublin.

O'Hare, B. (1982) *Maigh Eo, Mayo*. Galway.

O'Meara, J. (1982) *Topographia Hiberniae by Giraldus Cambrensis (c.1185)*. London.

O'Ruadhain M. (1954) 'Birds in Irish folklore' in *Acta XI Congressus Internationalis Ornithologici, Basel*. Dublin.

O'Suilleabhain, A. (1979) *The diary of Humphrey O'Sullivan (1827–35)* (transl. by T. de Bháldraithe RIA). Dublin.

O'Suilleabháin, S. (1942) *A handbook of Irish folklore*. Dublin.

O'Sullivan, H. (1827–35) *A handbook of Irish folklore*. Dublin. Diary: see O'Suilleabhain, H.

O'Tuama, S. & Kinsella, T. (1981) *Poems of the Dispossessed. An Duanaire*. Dublin.

Otway, C. (1827) *Sketches in Ireland*. Dublin.

Oulton, A.N. (1836) *Index to the statutes of Ireland (1310–1835)*. Dublin.

Parker R.D. (*c.*1845) *Ireland's birds* (illustrations of Ireland's birds reproduced later in book form by Blackstaff Press). Belfast.

Patten, C. (1906) *The aquatic birds of Britain and Ireland*. London.

Payne, R. (1589) *A brief description of Ireland*.

Payne-Gallwey, R. (1882) *The fowler in Ireland*. London.

Pennant, T. (1812) *British zoology*. London.

Perry, R. (1978) *Wildlife in Britain and Ireland*. London.

Petty, W. (1700) *A geographical description of ye kingdom of Ireland*. London.

Quammen, D. (1996) *The song of the Dodo*. New York.

Ray, J. (1674) *A collation … with catalogue of English birds*. London.

Rinker, D. (1986) *The ornithology of Eua*.

Ritchie, J. (1920) *The influence of man on animal life in Scotland*. Cambridge

Royal Irish Academy (1983) *The dictionary of the Irish language (old and middle Irish)* (compact edition ed. E.G. Quinn). Dublin.

Ruttledge, R.F. (1966) *Ireland's birds*. London.

Rutty, J. (1772) *An essay towards the natural history of the county of Dublin*. Dublin.

Salvin, F.H. & Brodrick, W. (1855) *Falconry of the British Isles*. London.

Sampson, G.V. (1802) *Statistical survey of Londonderry*. Dublin.

Saunders, H. & Clarke, W. Eagle (1927) *Manual of British birds*. London.

Seigne, J.W. (1918) *Irish bogs*. Dublin.

Sibbald, Sir R. (1684) *Scotia illustrata*. Edinburgh.

Simington, R.C. (ed.) (1962) *Books of survey and distribution (1636–1703)*. London.

Sleeman, P. (1989) *Stoats and Weasels. Polecats and Martens*. London

Smith, C. (1750–74) *The antient and present state … Cork, Waterford, Kerry*. Dublin.

St John, C. (1893) *Wild sports and natural history in the Highlands*. Edinburgh.

Stout, M. (1997) *The Irish ringfort*. Dublin.

Stringer, A. (1714) *The experienced huntsman* ed. by J. Fairley. Belfast.

Summers, D. (1976) *The great level – a history of drainage and land reclamation in the Fens*. London.

Thompson, W. (1849–51) *The natural history of Ireland*, 3 vols. London.

Turner, W. (1544) *Historia avium*. London.

Ussher, R.J. & Warren, R. (1900) *The birds of Ireland*. London.

Vincent, P. (1990) *The biogeography of the British Isles*. London & New York

Ware, J. (1654) *De Hibernia et antiquitatibus ejus*. Dublin.

Watson, J. (1997) *The Golden Eagle*. London.

Watters, J.J. (1853) *The natural history of the birds of Ireland*. Dublin.

Weld, C.P. (1857) *Vacations in Ireland*. London.

Whilde, T. (1977) *Birds of Galway and Mayo*. Galway.

—— (1993) *Threatened mammals, birds, amphibians and fish in Ireland* (*Irish Red Data Book 2: Vertebrates*). Belfast.

Whilde, W.R. (1861) *A catalogue of animal materials and bronze* (RIA). Dublin.

White, N.B. (ed.) (1932) *The Red Book of Ormond* (15th cent. transcript) Dublin.

Wilson, E.O. (1992) *The diversity of life*. London.

Willughby, F.J. (1676) *Ornithologica* (translated and enlarged by J. Ray, 1678) London

Wood–Martin, W.G. (1866) *The lake dwellers of Ireland*. Dublin & London.

Woodward, T., (ed.) (1729) *A collection of tracts concerning the present state of Ireland*. London.

Yapp, W.B. (1962) *Birds and woods*.

—— (1981) *Birds in medieval manuscripts*.

—— (1982) *Birds in captivity in the Middle Ages*. London.

Young, A. (1780) *A tour in Ireland (1776–1779)*, 2 vols. London.

JOURNALS, PERIODICALS & PAPERS

Investment in today's and tomorrow's environment, Dublin. An Foras Forbartha (Report) 1980.

'Naturalist's report by J. Templeton'. *Belfast Monthly Magazine* 1808–14: 10 vols (excl. vol. 2).

Bird Study (British Trust for Ornithology).

Boda Wennol (1998): *Newsletter of the Welsh Kite Trust*.

'Crackdown on Crane-killers'. BBC, *Wildlife Magazine*, May 1985.

British Birds (1907–99).

'Naive birds and noble savages – a review of man-caused prehistoric extinctions of island birds'. *Ecography* 16: pp 229–50. Copenhagen, 1993.

'A catalogue of the birds of Ireland by Dr P. Browne'. *Exshaw's Gentlemans and London Magazine*, June 1774, pp 358–87.

An Irish vegetational record, Littletown Bog, Tipperary by G.F. Mitchell. Geological Society of America, Special Paper 84, 1965.

'The Capercaillie' by C.E. Palmer. HMSO Publication no. 37 (1965)

'The former status of the Crane, Grus grus in Britain' by S. Boisseau & D.W. Yalden. *Ibis*, 140, pp 482–500.

Irish Birds.

Irish Bird Report (1953–76) Nos. 1–23.

Irish Geographical Studies (1970) 342–59, 'The history of land drainage in Ireland – Roscommon'.

Irish Naturalist (1892–1924) vols. 1–33.

Irish Naturalists Journal (1925–99).

Journal of Ardagh and Clonmacnoise Antiq. Soc. (1923), 25–31 and 'Description of Longford, 1682' by N. Dowdall (1932), 25.

Journal of the Dept. of Agriculture, 49, 'Origin and dev. of arterial drainage in Ireland and the Pioneers' by J. O'Loan.

Journal of Biological Conservation 9:1 (1976), 45–54, 'Towards a philosophy of nature conservation'.

Journal of the British Trust for Ornithology (Bird Study), 16 (1969), 193–209, 'The number of bird species on islands' by D. Lack.

—— 28 (1981) 33–41 'The food of the Red Kite in Wales' by P.E. & J.C. Davis.

—— 39 (1983): Pt.3 'Nightjars in Britain and Ireland in 1981' by F.C. Gribble.

—— 131 (1984): 'Raptors in Britain – a review of the last 150 years' by I. Newton.

Journal of the Cork Historical & Archaeological Soc. 98 (1984), 33–54.

Journal of Environmental Management, 8:2 (1972), 137–149, 'Post–war land drainage, fertilizer use and environmental impact in N. Ireland' by D.N. Wilcock.

Journal of the Galway Arch. & Hist. Soc. (1944–5) Vol. 21; (1952) Vol.25, No. 1, 4.

Journal of the Geological Soc. of Dublin, 1 (1833), 224–231, 'Notice of animals which have disappeared from Ireland' by Dr Scouler.

Journal of Irish Economic and Social History, 10 (1983), 51–65, 'Land use in medieval Ireland: a review of the documentary evidence' by H. Jager

Journal of Irish Forestry, 1:2 (1943); 38:2 (1981), 92–100.

Journal of the Royal Soc. of the Antiquaries of Ireland, 77 (1947), 63–80 'Irish cave excavations' by J.C. Coleman.

—— 79 (1949), 126–139 'Ballinagoola – Marshland habitation sites' by S.P. O'Riordain.

Journal of the Zoological Society.

Magazine of Natural History, (1831) 'Birds at or near Londonderry' by H.N.

—— (1832) 'A list of birds … on the northern coast of Donegal' by J.V. Stewart.

—— (1837) 'Selections from papers of J. Templeton' by R. Templeton.

National Board for Science and Technology, Dublin (1980) *Impacts of drainage in Ireland* (papers presented at a workshop).

National Wildlife Magazine (National Wildlife Federation, USA).

Proceedings of the Dublin Natural History Society.

Proceedings of the Royal Irish Academy (1836–).

Quaternary Science Review 16 (1997), 29–159, 'The Irish Quaternary fauna project' by P.C. Woodman, M. McCarth, N.T. Monaghan.

Report and Proceedings of the Belfast Naturalists Field Club (1886) Series 3, Vol. 2, 437–440.

Extinct wildlife by Revd J. Grainger.

New Scientist 68 (1975), 703–5, 'Saving wildlife for a rainy day' by R. Allen.

Scottish Birds: Journal of the Scottish Ornithologists Club.

Studies 20 (1931) 'The origin of Irish nature poems' by G. Murphy.

—— (1944), 249 *The Cock of the Woods* by Fr P.G. Kennedy SJ.

—— (1970) Autumn. *Conservation of the Environment in Ireland* by J.T. Lang.

Tauraco 2 (1992), 1–23. 'Status & conservation of Corncrakes Crex crex outside the breeding grounds' by T.J. Stowe & D. Becker.

Technology Ireland (1976), 'Wildlife – nuisance or asset in changing Ireland' by A.Whilde.

Transactions of the Institute of Civil Engineers of Ireland 103 (1978–79), 53, 'Land drainage and reclamation'.

Transactions of the Kilkenny Archaeological Soc. 2:1 (1852), 144–55, 'Of Hawks and Hounds in Ireland' by J.P. Prendergast.

—— (1870) 1:1, 139, 'Beranger's tour of Connaught'.

Transactions of the Royal Irish Academy 1–33 (1787–1907).

Ulster Folklife 11 (1965), 98–135; 12 (1966), 104–7; 17 (1971), 81–4, 'Local bird names in Ulster' by J. Braidwood.

Ulster Journal of Archaeology 18 (1055), 45–82, 'Excavations in Lough Faughan crannog, Co. Down' by A.E.P. Collins.

INDEX OF PLACES

INDEX OF PERSONAL NAMES